AF572672

STARS, MEN AND ATOMS

HEINZ HABER

Stars, Men and Atoms

GOLDEN PRESS NEW YORK

Library of Congress Catalog Card Number: 62-14287

First Printing, March 1962

Second Printing, 1963

 Designed and produced by Golden Press, Inc. Printed in the U.S.A. Published by Golden Press, Inc., New York. Published simultaneously in Canada by The Musson Book Company, Ltd., Toronto.

Table of Contents

1 • *Man and His Earth in the Universe*

What comes to mind when we hear the word "Earth"?

Some of us think of a globe of the kind we have sitting on our desk — a colored ball that shows the oceans in blue and the continents in their characteristic colors, one for each country. Or we think of a world map depicting the familiar outlines of the continents spread out on a large rectangle.

But to some other persons the word "Earth" means something quite different. It means a planet, one of the family of the solar system, a great ball that drifts through space against the backdrop of the universe sprinkled with its innumerable stars.

It is the fathers and mothers of today, the "older generation," who think of globes and maps. To youth, Earth is a celestial body, an astronomical subject like Venus and Mars. Young people envision the planet as it would be seen from space — as if already they were space travelers.

The mid-twentieth century is indeed a turning point in the history of how man has viewed the planet on which he lives. In this history there have been three phases, of which the first lasted longest. From the dawn of civilization to the early sixteenth century, man generally thought of himself as living on a flat disk. Then the first successful circumnavigation of the world, completed by Magellan's expedition in 1522, ended a long controversy, proving beyond doubt that our home is the surface of a sphere. Thus began the era of desk globes and maps. And now we are beginning the third phase, with our young people, especially, grasping this new concept of man's cosmic abode, of the position of Earth in the universe. It is youth who best demonstrate that we are now in the Age of Space, at a new lookout point on the long and tortuous road of human effort and understanding.

All through history, man has stubbornly clung to the conceit that he is at the center of the universe. In all the past of science and exploration, few human ideas have been disproved so often yet held so loyally. The process by which man has been dislodged, step by step, from his central position in the universe is a process of intellectual growth that we owe to many great minds of the past. And only in our own generation are we witnessing the final phases of this historic development.

The ancient peoples who believed Earth to be a huge disk also thought their own country was right in the middle. To the Chinese, China was always the "land of the middle," and a similar belief about their countries was held by the peoples of India and by the Mayas of ancient America. The precise Greeks even defined a mathematical center of the world disk: this center was located in Delphi, the famous sanctuary of Apollo. Delphi was on the slopes of Mount Parnassus, a few miles inland from the Gulf of Corinth. The temple was most sacred to the Greeks because of its famous Oracle, which was consulted by kings, statesmen, and private citizens who needed to know how their ventures were going to turn out. But the temple at Delphi had still another significance: in the middle of the spacious halls of the main building stood an adorned cone of highly polished marble, the "omphalos" — the navel — of the world.

It is easy to see why the human mind would imagine a disk-shaped world. Wherever we are, we always see ourselves as in the center, while the rest of the world is arranged in concentric circles around us. Out on the ocean, we see a vast circle of horizon around us. The night sky is a dome seemingly centered directly over us.

The early Greek thinkers pictured the earth as a broad, round expanse girdled by the river Okeanos. This picture did not conflict with the geographical knowledge of the Mediterranean peoples at that time. When, in later centuries, the Greek philosophers became dissatisfied with this primitive view, their new idea that the world is round was not prompted by new geographical discoveries but by their own idealized mathematical thinking.

We do not know who first had the tremendous idea that Earth is not a disk but a sphere. We do know that the famous Greek philosopher Aristotle, who lived from 384 to 322 B.C., believed it. He reasoned that it must be so because the sphere had the perfect mathematical shape, and only the perfect shape was fit to represent Earth and the universe. Not content with this argument alone — powerful and convincing though it was to the orderly Greek mind — Aristotle also rendered a most elegant proof. He pointed out that in a lunar eclipse, when the Moon moves into Earth's shadow, the edge of the shadow is circular. Earth's shadow as projected onto the Moon is *always* a segment of a circular disk, said Aristotle, and among all geo-

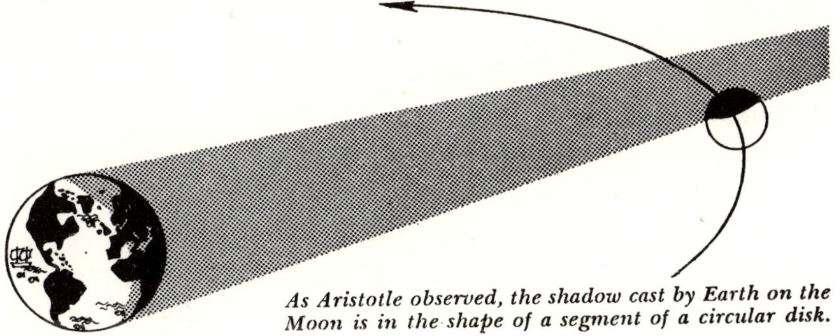

As Aristotle observed, the shadow cast by Earth on the Moon is in the shape of a segment of a circular disk.

metrical bodies only the sphere casts a circular shadow from any and all positions. This was a simple but most ingenious proof. It is indeed impressive to watch a lunar eclipse and realize how our planet presents its true shape on this cosmic screen provided by nature.

With the recognition that Earth is in truth a sphere, people lost their claim to a central position in the world. All locations on the surface of a sphere are mathematically equal in rank; not a single location is to be distinguished from any other. The claim to be in the center of the world belonged to all people at once — or to none.

The fact that Earth is a sphere did not become common knowledge until long after the time of Aristotle. The idea was not taught in school; rather, it remained the possession of only a few philosophers, astronomers, and mathematicians. It was all but lost during late antiquity and all through the Middle Ages, until in the fifteenth century it became an exciting speculation among the courageous navigators of Spain and Portugal.

The reason for this speculation was quite practical. Spain and Portugal were virtually cut off from the rest of Europe by a steep barrier of mountains, the Pyrenees. The Moslems dominated the countries of the Middle East, barring Europeans from trade for the riches of India. Blocked in those directions, the Portuguese and the Spanish probed around the continent of Africa and wondered whether it would be possible to reach India by sailing westward around the spherical (so it was hoped) globe. The true shape of Earth was, then, of great practical importance, while before that time it had remained nothing more than an interesting mathematical and astronomical speculation.

But even at the time of Columbus, the spherical shape of Earth was denied by many intelligent people. In his pleas to the court of Spain for support of his plan to reach India via a western route, Columbus ran against the classic argument: "What will happen to your ships when you reach the end of the world? Will they not slip over the rim and plunge into nothingness?" Only after long years did Columbus win his point and set sail.

To the end of his days, Columbus was convinced that he had actually reached the eastern coasts of India, and that he had truly crossed the unknown portion of a round Earth. Of course, that was not the case. First to complete the circumnavigation of the globe was the expedition of the Portuguese navigator Ferdinand Magellan. On September 20, 1519, he set out from Sanlúcar de Barrameda, the harbor of Seville in Spain, with five ships. He went around the southern tip of the Americas, forged across the broad wastes of the Pacific, and after desperate hardships reached the Mariana Islands and discovered the Philippines. While assisting a native chieftain in war against a neighboring island, Magellan was killed in 1521. One of his remaining ships, the "Victoria," under the command of Captain del Cano, completed the world trip by rounding Africa and arriving again in Seville on September 9, 1522, almost exactly three years after the start. Oddly enough, the ship's calendar gave September 8 as the date of arrival; nobody gave thought to the fact that a day is lost on a westward circumnavigation of Earth!

After the feat of Magellan, the true shape of the Earth became commonly accepted knowledge. The second phase of man's view of the planet had begun.

No one, now, could rightly claim to occupy a central position in the world. But Earth itself was still believed to be the center of the universe, with all the stars, Sun, Moon, and planets revolving around it. That notion, too, was doomed, and the man who was destined to destroy it was already forty-nine years old when the "Victoria" arrived in Seville after her historic voyage.

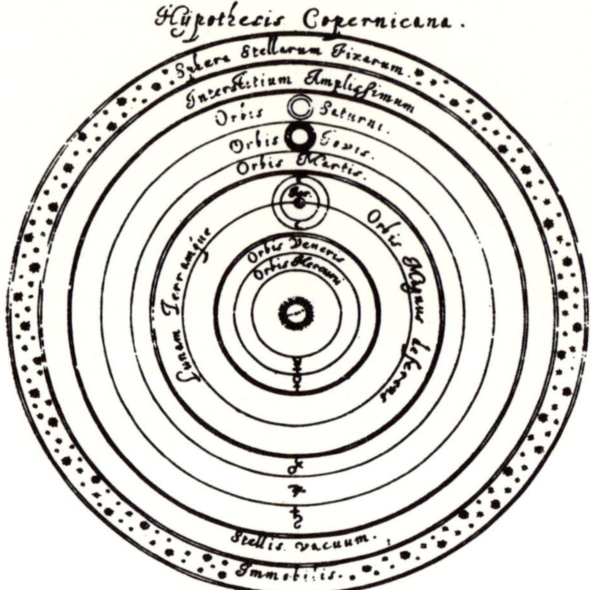

Copernicus' revolutionary theory of the universe envisioned the Sun surrounded by the orbiting planets. He went wrong in the outermost layer, which he thought to be a single sphere holding all the fixed stars.

It was in 1543, the year Nicolaus Copernicus died, that his famous book *De Revolutionibus* was published. It assigned the central position of the universe to the Sun – which is, after all, entitled to this place owing to its overwhelming brilliance. Earth was demoted by Copernicus to a humble station among the planets that circle the Sun, far from the coveted place in the middle.

When the ideas of Copernicus became known, most learned men – like everyone else – flatly refused to believe. Could this mighty Earth, the firm ground under their feet, together with the vast oceans and the air, be moving through the sky? Copernicus, the "Polish fool," said Earth was swinging about its axis once every twenty-four hours, and at the same time flying around the Sun in a gigantic motion more than a hundred times faster than a cannon ball!

A century and a half had to pass before most educated people would accept this grotesque idea. The meager power of human visualization had not been the only obstacle. The human mind was also reluctant to accept the humbling fact that Earth, the home of man, is not the center of creation.

Yet the Copernican revolution was only the beginning. Once Earth, and with it man, were flung out into space, other new ideas were set in motion. The book of Copernicus had treated only the planets and their motions; the stars were still looked upon as luminous points fastened to a huge sphere that enveloped the entire universe, with the Sun in the center. For thousands of years people had believed in the existence of this crystal sphere that bore the stars; do we not see it each clear night all over the world? Yet after Copernicus the days of the stellar sphere, too, were numbered.

A few decades after the great Pole's death, the Italian monk Giordano Bruno was calling the stellar sphere an illusion. He taught that the "fixed stars" are in truth suns — suns that rival our own in size, majesty, and brilliance. He recognized that the universe is infinitely greater than the little space occupied by the solar system. He shattered the idea of the stellar sphere and opened up views of a vast space so wide, high, and deep that it afforded room for millions of solar systems. So great are the distances between us and these suns, he thought, that their overpowering glare is dimmed to the weak twinkle of stars seen at night.

In the year 1600 Giordano Bruno, declared a heretic, was burned at the stake. The story of his trial is involved; he was found guilty not because of his astronomical theories alone. But the solemn priests of the Inquisition were, to say the least, profoundly shocked by this man who could envision Earth as a tiny, insignificant ball floating through space like a bit of dust. In their eyes such an idea was sheer blasphemy. Had not Earth and the human race been favored by the coming of the Savior? They *must* be more important than the "mad" monk thought.

Even today, there are many persons who cannot make the modern scientific view of the universe fit their religious beliefs. But within two hundred years after the death of Giordano Bruno, most scientific-minded thinkers had got used to the idea that Earth is but a speck in the galaxy. The universe was pictured by astronomers as a huge, wheel-shaped cloud of stars, visible to us inside it, as we look toward the edges, as the silvery band of the Milky Way.

The English astronomer William Herschel, probing the galaxy with his powerful telescopes, decided that the Sun with its planets is located near the center of the Milky Way. So, while Earth had lost some distinction, man at least found himself still in the middle of the known universe. For there was only one galaxy — namely, his own.

About the time of Herschel, a little less than two centuries ago, astronomers were wondering about the nature of certain faintly luminous patches in the heavens that could not be resolved, or separated, into individual stars. As larger and better telescopes were built, more and more of these luminous patches, called "nebulae," were found. The German philosopher Immanuel Kant was the first to suppose that these faint clouds were other star systems — other galaxies at awesome distances from ours. But there were few willing to listen. The idea was too ridiculous.

In the nineteenth century the distances between the stars became more accurately known, and the position of the Sun within our galaxy was re-examined in terms of this new knowledge. The final result was a further blow to the old conceit that man is in the center of things. The Sun was found to be much closer to the rim of the Milky Way than to the center. And growing knowledge of the Sun identified it as just a modest, average star, distinguished from millions of others neither by size nor by position.

In 1917 the 100-inch telescope, then the largest telescope of the world, went into operation at the Mount Wilson Observatory in California. Among its first targets were the faint nebulae. A few years later there was proof beyond doubt that these mysterious pale patches of light are indeed galaxies, equal in rank to our own. At first it looked as though our galaxy was a giant compared to the average run; perhaps we still had something to brag about. But that belief, too, was shattered after World War II, when Dr. Walter Baade of the Palomar Mountain observatory discovered an important error in the accepted dimensions for the universe. Today we know that our own Milky Way galaxy is of an ordinary dime-a-dozen sort; there are billions equal in size or even bigger.

Now, indeed, man and his tiny Earth had shrunk to insignificance. Yet he still clung to another distinction — one that goes beyond the mere geometry of the universe. Was not mankind probably the only intelligent race in the universe? Not long ago, this view seemed well founded; it was believed that our solar system was created by a celestial event so rare and improbable that the chances for the existence of other life-supporting solar systems in the universe were negligible. Only thirty years ago the English physicist Sir Arthur Eddington, a foremost authority, remarked that life on our planet must be unique in the universe. He looked upon the existence of life on Earth as the result of an oversight, as it were — as the result of a failure of nature to keep the universe "sterile."

In the past decade, this notion of man's solitary glory has gone the way of similar ideas of the past. Today science is fairly certain that among the billions of alien suns in the depths of the galaxies many are adorned with families of planets. Of course, there are many stars unlike our Sun. Perhaps most planets are as unfriendly to life as Mercury, for example, or Jupiter — too hot or too cold. Most may be sterile, because nature has denied them the physics and chemistry that would make an evolution of life possible. Yet if the chance is but one in a million, there are certainly thousands, even millions, of other planets teeming with life under the warmth of friendly suns. And why should not these living beings, too, be blessed with what we call the spark of intelligence?

This, then, is a lesson out of more than two thousand years of search and wonderment: man is not in the middle of, and he is not unique in, the universe.

The macrocosm, however, the magnificent realm of the stars and galaxies, is only one aspect of creation. There is the microcosm — the realm of molecules, atoms, nuclei, elementary particles. What is the position of man between these two major divisions of nature?

The human body, like any other physical object, consists of atoms. It takes a tremendous number of atoms to build the body of a man, because the individual atoms are so very small. The number of atoms required to build the body of a man is so great that it doesn't have an ordinary name; it is a number with 28 zeros. Call it n. Now, there is an interesting possibility in that number n. What would we get if we took a mass consisting of n human bodies? The sum of n human bodies makes a mass that weighs about as much as an average star! So the size of man is in the middle between atom and star. As Eddington first pointed out — in the equal ratios of atom, man, and star —

$$\frac{\text{Atom}}{\text{Man}} = \frac{\text{Man}}{\text{Star}}$$

Man occupies the middle register.

Now, the atom is not the smallest thing in nature. It is composed of much smaller particles, such as protons, neutrons, and electrons. And the average star is by no means the largest thing in nature. The Milky Way and the other galaxies in the universe are much bigger. We choose the atom and the star for our subjects because they are two of the most fundamental and distinct formations in nature. They are, in a sense, the basic units of creation.

To find man in the middle between atom and star is a sort of trick – certainly not an important scientific discovery. But it makes one ponder. Standing as he does between the microcosm and the macrocosm, man is in a highly convenient position. He can readily explore and even understand the miracles of creation in both directions. If modern science has robbed him of the age-old dream of uniqueness, of being in the geometrical center of things, it has given him something more valuable instead. What a stupendous spectacle is this universe of ours, with man so small yet so marvelously able to explore in all directions, to comprehend, and to pass along his knowledge. The search for understanding, for the truth of our position in creation, is no less wonderful – and indeed it is far more rewarding – than blind clinging to the egocentric dreams of the past. Man has not been made smaller by knowledge, but larger. Man thinking, man seeking to understand, wherever he may be, is truly at the center of things.

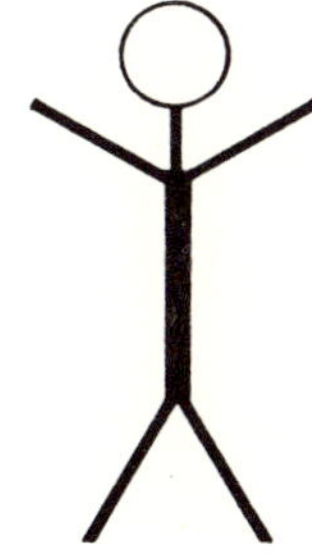

A planet drifting through space, a universe of suns and planets, a range of natural laws from galaxy to atom – this is the modern view of man's world. The view was hard-won, and the older heads among us still tend to think in terms of the globe on the table, the map on the wall. It is the young who know the Earth as a planet, swinging down its age-old path against the backdrop of the universe. Full of plans for space travel, for exploring the depths of the atom and the far-flung chains of galaxies alike, the young are not dismayed. They are – if their parents are not – in the middle, or perhaps we should say at the center, of things after all.

2 • *Fantasy and Logic*

To man the sky was once a huge dome fashioned of the purest crystal, rendered blue by the light of the day, and left black during the night. At night this black dome was adorned with innumerable sharp, brilliant points – the stars, fixed upon the crystal in never-changing stations.

In the beginning man explained the stars in terms of poetry and religion. The apparent motions of the stars were followed for perhaps a hundred centuries before there was any understanding their true nature. The stars must have been among the very first objects in nature that started man to thinking thoughts beyond the daily business of survival. Contemplation of the universe made early man, at least in some of his moments, a storyteller, a poet, and a philosopher.

Astronomy has justly been called the oldest of the sciences, but its age makes difficult any tracing of its beginnings. The most ancient religious myths and poetic fables about the stars were often based on keen observations that could almost be called scientific. More often than not, however, observed astronomical facts were twisted to satisfy the storyteller's or poet's sense of drama.

Talk though they may of old-time superstitions, many people today have never known the full impact the starry sky has upon the human soul. Ancient man sat in the starlight; we sit in the light of table lamps. The dust and smog of cities obscures the brilliance of the stars, or the light of these other suns is outshone by the light bulbs along our streets. How many of us know how splendid the stars truly are? We forget the night sky as it is in all its undimmed pageantry. Then, during a ski trip in the high mountains, or at night on the sea, the grand spectacle unfolds. It is a shock to realize that it has been there all the time.

Primitive peoples are great stargazers. The splendor of starry skies is an intimate, important part of their lives. Peoples everywhere have grouped together certain conspicuous stars and have seen in these constellations all kinds of human beings, gods, animals, and everyday objects. Because the constellations change their forms or relative position only very slightly during thousands of years, they seem eternal and remote from poor man — and appeal the more to his poetic and religious imagination. So the folklore of all peoples contains treasures of magnificent fables and myths of how the various star figures came into being. Studying these myths through the eyes of modern science is especially interesting because the story material of many fables reveals a keen observation of astronomical facts.

As the rotating Earth revolves around the Sun, it carries us around the inside of the spherical stage of the sky. Of course, we are not aware of this constant motion of our planet; rather, the whole sphere of the heavens appears to rotate about our station once every day. But we are made aware of Earth's progress in its journey around the Sun by the fact that at a given moment each night the parade of constellations is slightly westward of the point where it was the night before. So, in the course of a year, the complete circle of the heavens is completed.

Earth rotates on an axis whose ends are called the poles. If the north pole were actually a very long pole, it would seem to pierce the dome of the sky. The point at which that would occur is called the north celestial pole. Very near it, within half a degree, is the star called Polaris, the north star. Polaris seems to remain stationary while the other stars, because of earth's rotation, seem to revolve around it.

The one star that seems to stand still in the heavens becomes a perfect subject for legends. To the Chinese the polar star was the emperor of the heavens, who sat upon his immovable throne while the other stars paid homage to him. In another Chinese myth the polar star was the goddess of wisdom, placed in the sky and deified because she was a wise and virtuous woman all her life. The practical Phoenicians, bold navigators of antiquity, were the first to recognize the great importance of the north star as an aid to navigation; for with its immovable station it always indicated the direction of north. The Phoenicians called it the Lodestar, the Ship's Star, or the Star of the Sea.

The polar star does have an apparent motion; it does not stay at the exact mathematical point that is the north pole of the sky. Once every day it traces a small circle about twice as big as the disk of the Moon.

Due to a "wobble" in the Earth's axis, a line drawn through the poles and extended upward from the north pole describes a circle in the sky, making a full rotation about once every 26,000 years.

Over the centuries, the distance of Polaris from the exact north celestial pole is changing slowly, because the heavenly poles themselves are drifting. This drift is caused by a motion of Earth itself. Like a slowly spinning top, Earth has a wobble, called "precession." It is a very, very slow motion; the Earth's north pole takes almost 26,000 years to complete one whole circuit in its wobble. But in this time our imaginary pole traces an enormous circle on the sky dome. Obviously, then, a few thousand years hence Polaris will be nowhere near the celestial pole. In about 11,500 years the axis of the earth will point at the star Vega in the constellation of Lyra, the Lyre. Then we will have a particularly bright polar star, because Vega is the brightest star of the northern skies.

The polar star of today is not the polar star that was watched by the pyramid-builders in Egypt nearly 4,000 years ago. The polar star of their ancient stories was Thuban, a star in the constellation of Draco, the Dragon. It was for good reason that the figure of the dragon played an important, even a dominant, role in ancient mythology.

Among the most famous star figures, and perhaps some of the oldest ones, are the constellations of the Zodiac. There are twelve in all: Aries, the Ram; Taurus, the Bull; Gemini, the Twins; Cancer, the Crab; Leo, the Lion; Virgo, the Virgin; Libra, the Scales; Scorpius, the Scorpion; Sagittarius, the Archer; Capricornus, the Goatfish; Aquarius, the Waterman; and Pisces, the Fishes. These constellations form a broad band that rings the entire sphere of the heavens. They are particularly distinguished from all other constellations because this band is the apparent annual pathway of the Sun. The Sun, of course, remains at the center of the planetary system, and its apparent motion in front of the stars is due simply to Earth's yearly revolution about the Sun. As Earth travels along its huge orbit, our line of sight to the Sun shifts slowly, so that the Sun itself appears to drift in front of the stars. Each month the Sun is seen against a different constellation of the Zodiac; there is one constellation for each month of the year.

The orbits of Moon and planets all lie nearly in the plane of Earth's orbit. So, when seen from the Earth, these heavenly bodies, too, always appear against constellations of the Zodiac. In short, the ring of the Zodiac is the main highway of Sun, Moon, and planets, and that is why these twelve constellations are so famous.

Since the Sun appears to circle the Zodiac annually, the Zodiac is linked with the cycle of the seasons. The signs of the Zodiac derive their symbolic meaning from the yearly seasonal cycle. In spring the Sun is seen in the constellations of Aries and Taurus, the symbols of fertility. In summer, when the Sun displays all the fierceness of its heat, it finds itself in the Lion. A month later it is harvest time, and so the Virgin is shown holding in her hand the bright star "Spica," which is Latin for "ear of wheat." Later, during what was the rainy season in the world of the Mediterranean and eastward, the Sun travels through the "watery" constellations – Capricornus, Aquarius, and Pisces.

The mathematical track of the Sun's apparent motion through the band of the Zodiac is a great circle that spans the whole sky; it is called the ecliptic circle, or ecliptic. It is, in fact, involved in the phenomenon of eclipses. From ancient Babylon comes an old story about the origin of solar and lunar eclipses that also involves the famous character of the Dragon. This Babylonian dragon – his name was Tiamat – was coiled around the ecliptic. He hated the Sun and the Moon, and when during their wanderings these heavenly lights came close to the dragon, he would try to swallow them or to lash them with his powerful tail. But this could happen only during two periods within a year, six months apart. At all other times the Sun and the Moon would succeed in avoiding the angry beast.

This story is based on a number of acute observations about the possible occurence of eclipses. The orbit of the Moon is tilted at an angle of about 5 degrees from the circle of the ecliptic, or yearly track of the Sun around the sky. The orbits of Sun and Moon are thus like two concentric hoops, one of them tilted a little with respect to the other. The hoops touch at two opposing points; one half of the lunar orbit lies above the ecliptic, and the other half lies below. The points of contact are called the "Moon nodes."

Now we can see why there is not always a solar eclipse when the Moon comes between Earth and Sun, which happens once every month. Most of the times, the Moon passes below or above the Sun as we see it. For a solar eclipse to occur, it is necessary for the Sun to be at or at least near one of the nodes so that the Moon can meet and cover it.

The shadow of Earth lies on the ecliptic at a point exactly opposite the Sun. A lunar eclipse can take place only when Earth's shadow is on or very near one of the nodes. Only then will the Moon travel through Earth's shadow and be eclipsed; at all other times it will miss the shadow. But only twice during each year does the Sun pass through one of the nodes while the shadow of Earth is at the opposite node.

Now the story of the dragon Tiamat becomes clear: the beast's head was at one of the nodes, and its tail was at the other. To this day, the nodes of the Moon are called "the head of the dragon" and "the tail of the dragon." There are even symbols for these mathematical points, namely ☊ for the head, and ☋ for the tail. The dragon Tiamat still lingers in modern textbooks of celestial mechanics, where we can find equations such as

$$k_2 = \tan i \sin ☊$$

which reads: "*k*-sub-two equals tangent-*i* times sine of the head of the dragon"!

Another legend that involved celestial phenomena of great interest to modern astronomers was that of the Princess and the Whale. It includes an ancient king of Ethiopia, Cepheus; his beautiful wife Cassiopeia; their lovely daughter Andromeda; a young prince named Perseus; and a terrible sea monster, Cetus the Whale.

The lovely Cassiopeia was terribly conceited. One day she went so far as to claim that her beauty even surpassed the charms of the Nereids, the princesses of the sea. The latter ladies took offense and complained to their father Poseidon, god of the sea. To punish the vain queen, Poseidon thrust his trident into the waves and thus created Cetus, the monstrous whale. Poseidon ordered Cetus to lay waste the coastal regions of Ethiopia.

In desperation, King Cepheus consulted an oracle, which pronounced that the terrible ire of Poseidon could be placated only by a supreme sacrifice: the beautiful Princess Andromeda must be offered to the monster. Chained to a cliff at the shore, Andromeda was awaiting her terrible fate when Perseus appeared. Perseus, as it happened, liked the looks of Andromeda, and he was well equipped to deal with the emergency.

In a previous exploit, Perseus had slain the famous Medusa. This unfortunate character had once boasted that her golden crown of hair was more beautiful than the goddess Athene's, and in retaliation Athene had changed the hair into a nest of writhing snakes — so horrible a sight that whoever beheld it was instantly turned to stone. Perseus had beheaded Medusa in her sleep, watching her reflection in his polished shield. Now, armed with Medusa's dreadful head, he was ready for all comers.

When the sea monster broke the surface of the sea to claim poor Andromeda, it made the mistake of viewing the object held high by Perseus. Turned to stone, the monster sank to the bottom of the sea. Perseus cut the fair captive's chains and led her back to the overjoyed king and queen.

Poseidon, however, was still annoyed with the vain Cassiopeia. The noble family was transported into the sky to serve as an eternal warning to all men, and Cassiopeia was put into a chair. There, as seen by sky watchers in the northern hemisphere, she revolves around the north star once a day, upside down for half of the trip.

The head of Medusa, placed in the constellation Perseus, is represented by the star Algol. The name is from the Arabic *el ghoul,* meaning "demon," and the star does seem sinister. In every period of about 70 hours, there is an interval of about 10 hours during which Algol's brightness fades to a third of its maximum, then is gradually restored again. A winking star! No wonder it made a deep impression upon ancient sky-gazers.

Modern astronomy explains this strange fluctuation in Algol's brightness. Algol is in reality a double star — two suns that are revolving around a common center. When the dimmer of the

The two parts of the double star Algol rotate around each other about once every sixty-nine and a half hours. The chart shows the intensity of the light given off when the two come together (center) and when they are apart (on the sides).

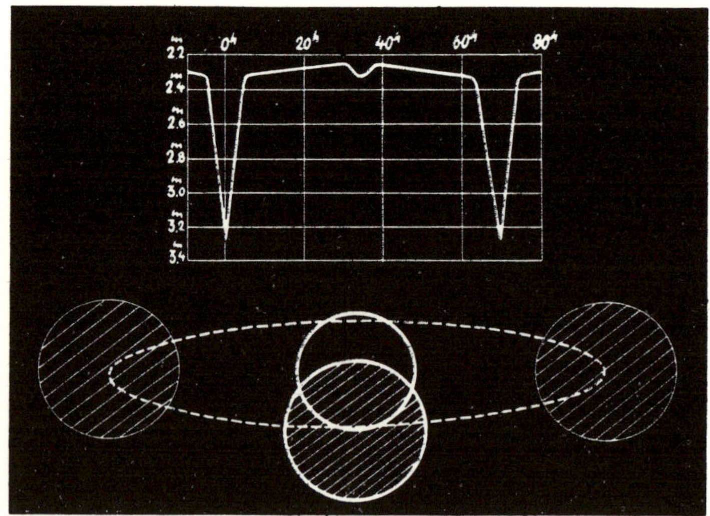

two passes in front of the brighter one, it cuts off some of the latter's light. This stellar eclipse, lasting 9 hours and 45 minutes, repeats itself regularly as the two stars revolve like clockwork. Although the modern astronomer cannot see the stars separately, he knows there are two because he is able to distinguish two different spectra.

With modern telescopes many of these so-called "eclipsing binaries" have been found, but Algol is by far the brightest. The ancient Greeks, so richly peopling the heavens, could hardly have found a star better fitted to represent the evil head of the unfortunate Medusa.

The name of the mythical Princess Andromeda is a name constantly used by astronomers today. In the beautiful chain of stars forming this constellation there is the brightest of the so-called "spiral nebulae" — those faint patches of light that defied explanation until the Mount Wilson telescope resolved them into individual suns. The Andromeda nebula was indeed the first to be resolved into a cloud of single stars by the powerful modern telescopes, and its nature as a galaxy similar to our own — though larger — was one of the most exciting discoveries of modern science, one that tremendously widened the known size of the universe in which we live.

If you live in the northern hemisphere, some night between autumn and spring when you are away from the lights of the big city, you may raise your eyes to the dark sky and see, faintly, that little patch of light in the constellation Andromeda. This is another universe, perhaps containing worlds like our own. Your modern knowledge is far from the legend of the beautiful chained maiden, daughter of Cassiopeia and Cepheus. But your wonder at the splendor of the heavens will be the wonder of the Greeks who viewed it over two thousand years ago.

For millennia, man viewed the sky with wild imagination and religious awe. It was impossible for ancient peoples to "explain" the vistas of nature as we moderns explain these things, scientifically. Today we demand to have the phenomena of nature explained and understood on the basis of known laws of nature that prevail everywhere. To the peoples of old there was no such thing as a natural law; everything was explained as soon as a god or demon was made responsible for whatever happened. Even in their first earnest attempts at being "scientific," thinkers could not help linking their ideas and their concepts with superstition and fantasy.

The beginning of scientific thinking goes back, perhaps, to man's first attempts to use the heavens as a clock and a calendar. He realized that the motion of the Sun created the day and the year, the Moon measured the week and the month, and the stars indicated the seasons. However, the heavenly reckoners of time were looked upon as gods similar to all other forces of nature that were beyond human ken.

Only in the sixth century B.C. did man break with random superstition and approach the contemplation of nature with a truly scientific attitude. This change came almost overnight through the systematic thoughts of the ancient Greek philosophers. Those enlightened men, actually the founders of scientific thinking, did more to change the course of human history than kings, statesmen, and soldiers. More history has been made by the keen weapons of the mind than by the crude weapons of war. The early Greek thinkers, more than 2,000 years ago, invented the method of scientific thinking and, above all, laid the foundations of mathematics.

These ancient Greeks who were such clear thinkers and accomplished mathematicians: why did they not develop a more advanced technology? They had all the necessary intellectual tools; they had developed a form of mathematics, which is still valid, practically, today. The mathematics taught to our high-school students today contains little that the ancient Greeks did not know, and much of Greek advanced mathematics even goes beyond our ordinary high-school curriculum. In Greece, wood- and metal-working were advanced enough so that it would have been only a short step to the invention of the steam engine or even the basic tools of electric power. As early as 465 B.C., the Greek philosopher Democritus devised the first atomic theory. Why, then, did the Greeks fail to go one step further and develop at least a basic science of chemistry?

The answers to these historical questions are found in the character of Greek scientific thinking. The great thinkers looked upon themselves primarily as philosophers in the precise sense of the word: "lovers of wisdom." Gadgets and other practical applications of their noble ideas they would have considered as an insult to the lofty spirit of science — certainly beneath their dignity as philosophers. A machine is essentially a labor-saving device, but any kind of manual occupation was left to slaves. A thinker doesn't need a machine.

The Greek scholars might have looked differently at slide rules, electronic computers, typewriters, and printing presses. But such things are the fruits of several centuries of technological development. Among these aristocratic intellectuals — these ancient "eggheads" — there was more satisfaction in finding proof of a tricky mathematical theorem than in the invention of some gadget for saving a slave's time.

It is in keeping with this spirit that the Hellenistic Greeks were the first to found a scientific library and an academy of sciences that could be called a university. Both were created in Alexandria, in ancient Egypt, in the early third century B.C. The most brilliant minds of that time were gathered there, making Alexandria the greatest intellectual center of antiquity. Among the most important subjects studied and taught at the university were astronomy and history, yet the most celebrated minds of the academy were mathematicians.

Alexandria was founded by and named after Alexander the Great, who conquered Egypt and left it under the command of one of his generals. In 305 B.C., eighteen years after Alexander's death, this general declared himself king of Egypt and assumed the name of King Ptolemy I "Soter." From him Cleopatra, the beautiful queen of Caesar's time, was descended. Under his wise reign Alexandria soon became beautiful and prosperous, and a focal point of commerce in the ancient world. Ptolemy, with his tremendous admiration for the learned mind, founded a library with the intent of collecting all known writings of his time. To enrich the collection he required all merchants entering the city to donate a scroll to the library, and since Alexandria was an extremely busy and profitable trading center, greedy merchants who were themselves indifferent to books, combed the entire known world for every kind of written material to use as tickets of admission. The library soon surpassed 600,000 scrolls; yet Ptolemy knew that no library is any good unless it is used, and so he founded the academy and invited to it the greatest minds of antiquity. With lifelong tenure at this first "university," the faculty members – mostly Greek scholars – were delighted to spend their lives with no obligation but to think.

The first librarian of Alexandria, who was also president of the university, was the famous mathematician Euclid, founder of geometry. His book on the subject is the only textbook of that age which is still used today in modern schools – without any major change. Geometry as it is studied today was laid out by this thinker of twenty-two centuries ago.

The successor to Euclid as chief librarian and president of the university was the great Eratosthenes. Astronomer, mathematician, and geographer, Eratosthenes won renown for his astonishing feat of measuring the size of the Earth. This was done in the third century B.C., at a time when few people in the world even suspected that the Earth was a sphere. The story is worth telling because it demonstrates a perfect, if simple, piece of scientific thinking.

In the Egyptian town of Syene, about 560 miles south of Alexandria, there was a deep well. Eratosthenes knew that in Syene, at noon of the longest day of the year, the Sun was exactly overhead. At this moment the Sun shone straight down the deep shaft of the well and mirrored itself in the water at the bottom. Eratosthenes, being a mathematician who knew something about angles and circles, and who believed the Earth to be a globe, thought about this well and made his plan. On the same day and hour that the Sun was overhead in Syene, Eratosthenes measured the length of the shadow cast by an obelisk, or needle-like monument, in Alexandria. Knowing the length of the shadow and the height of the obelisk, which also had been determined, the Greek genius knew enough to compute the size of the Earth!

The diagram on this page shows how simple it was. We can

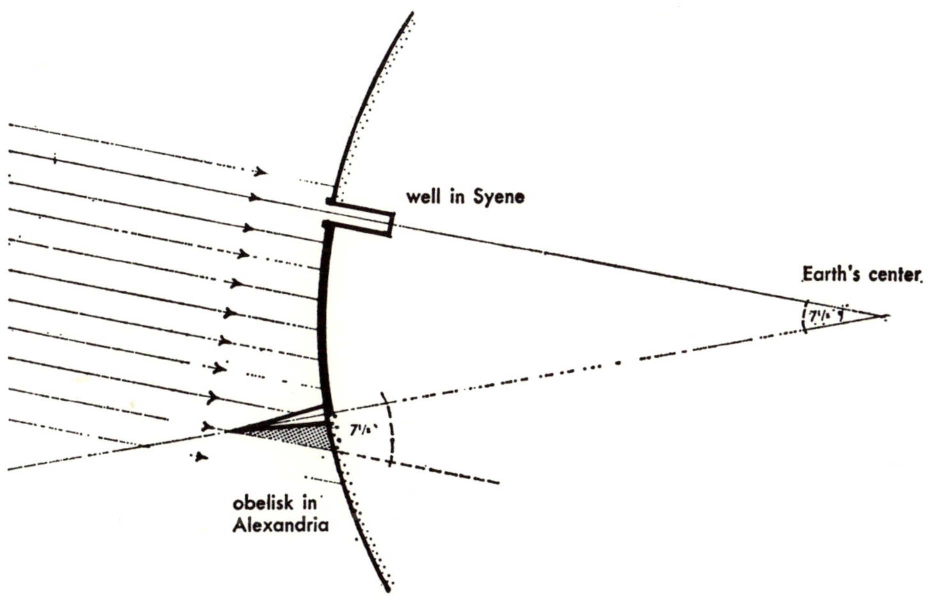

assume that Eratosthenes drew a picture of the obelisk and its shadow. He measured the angle between the obelisk and a line drawn from the top of the obelisk to the tip of the shadow on the ground. This angle was 7⅕ degrees. Now, in Syene the Sun was casting *no* shadow; the corresponding angle would be 0 degrees. Therefore, the distance of 560 miles between Syene and Alexandria was equivalent to 7⅕ degrees of Earth's circumference—assuming Earth to be a sphere. What would be equivalent to 360 degrees — Earth's entire circumference? Eratosthenes divided 7⅕ into 360, got 50, multiplied this by 560, and got a result of 28,000 miles for the total distance around Earth at its middle, the Equator.

Since the distance between Syene and Alexandria was not accurately known, the result obtained by Eratosthenes was 3,000 miles greater than the approximate 25,000 miles we accept today. But the error does not diminish in the least the intellectual feat of Eratosthenes. By the sheer power of mind he measured the size of the planet at a time when only a very small part of it — only the Mediterranean area — was well known. Eratosthenes foreshadowed the scientific achievements of thousands of years to come.

A famous story of the university of Alexandria does, however, show the utter contempt of these brilliant Greeks for practical applications of scientific knowledge. The story involves the man who was perhaps the greatest mathematician of antiquity — Archimedes, the "arch ponderer." Eratosthenes, much impressed by the mathematical genius of Archimedes, invited him to leave his home in Syracuse, a Greek colony on the island of Sicily, and come to Alexandria to join the faculty of the university. The great Syracusan accepted. Exactly what happened to him in Alexandria is not known, but the account we have at least expresses the attitude of a great many of the notable scholars who held sway there.

The disk-shaped shadow of the Moon as projected onto the Sun in this eclipse shows the Moon to be a spherical body. Likewise, in a Lunar eclipse, as Aristotle pointed out, the Earth's shadow on the Moon takes the same shape — proof that Earth, too, is round

A time exposure shows the march of the stars across the heavens, a spectacle that has enthralled man from his earliest days — even though it took many centuries before he began to understand what was happening

A view toward the center of our galaxy — the Milky Way. Prior to the teachings of Giordano Bruno (1548-1600) the "fixed stars" were thought to be imbedded in a "stellar sphere." Bruno recognized that each star was, in fact, a sun, and that there were countless numbers of them

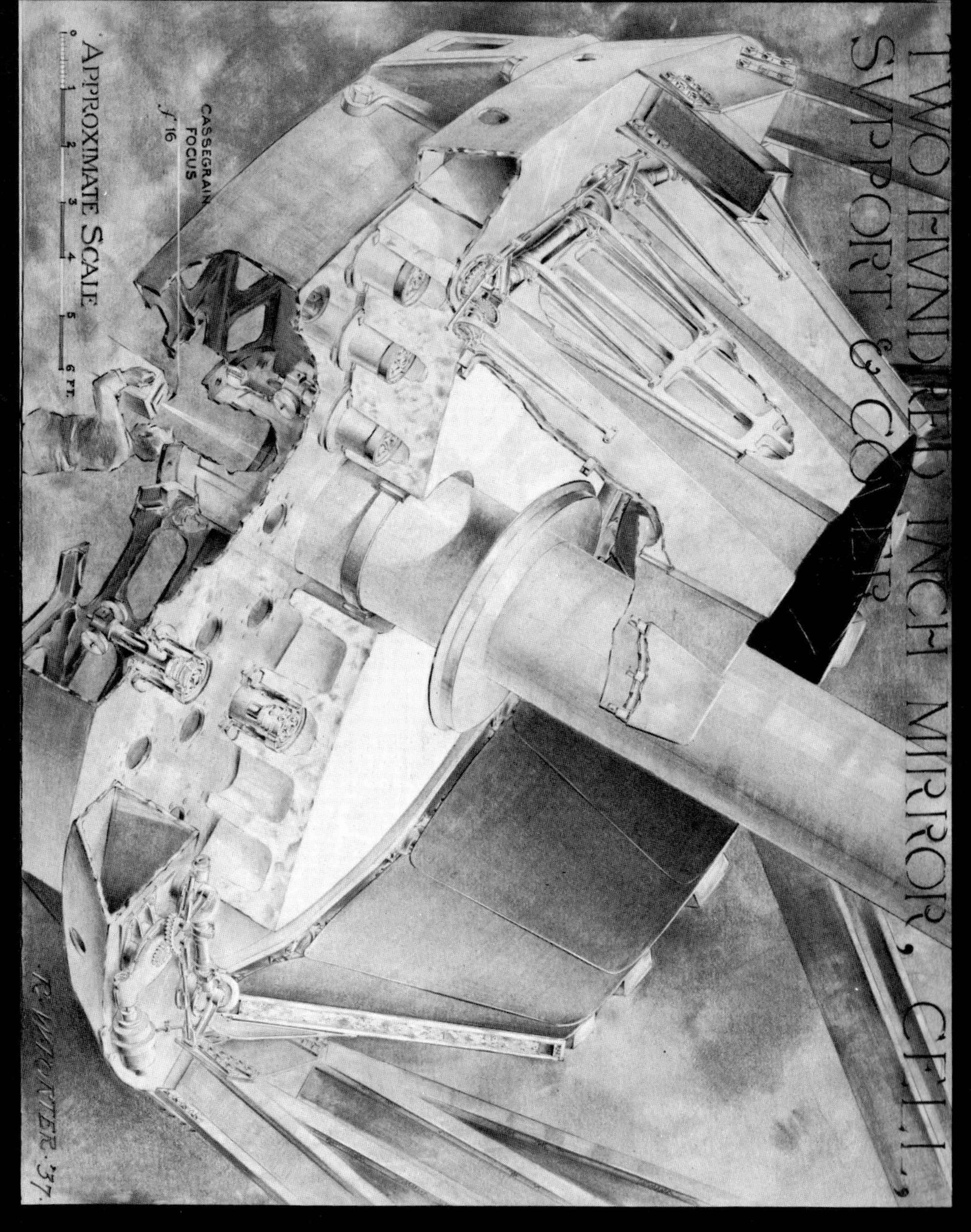

Russell Porter's drawing shows the Palomar observatory's 200-inch telescope. Through development of instruments such as this, the advances of modern astronomy were brought about

A spiral nebula — or galaxy — seen edge on. Roughly the equivalent of our own Milky Way, this star cluster may well include suns whose planets are so constituted that they support life — perhaps even intelligent life

This time exposure shows how stars seem to rotate around the polar star. Always keeping its same relative position, the polar star is of incalculable value to mariners

The famous spiral nebula M31 in the constellation Andromeda was the first nebula to be resolved into individual stars by modern telescopes

The wind blows snow across the Great Ross Ice Barrier in Antarctica. The ice here may sometimes be as much as two miles thick, yet in proportion to the rest of the Earth, it is but a touch of frost

Lt. Colonel John H. Glenn, Jr. observes as a technician checks over the "Friendship Seven," the space capsule that had just returned him safely to Earth after his successful four-hour ride through space

Russia's Major Yuri Gagarin, the first man to travel out into space and come back to tell about it

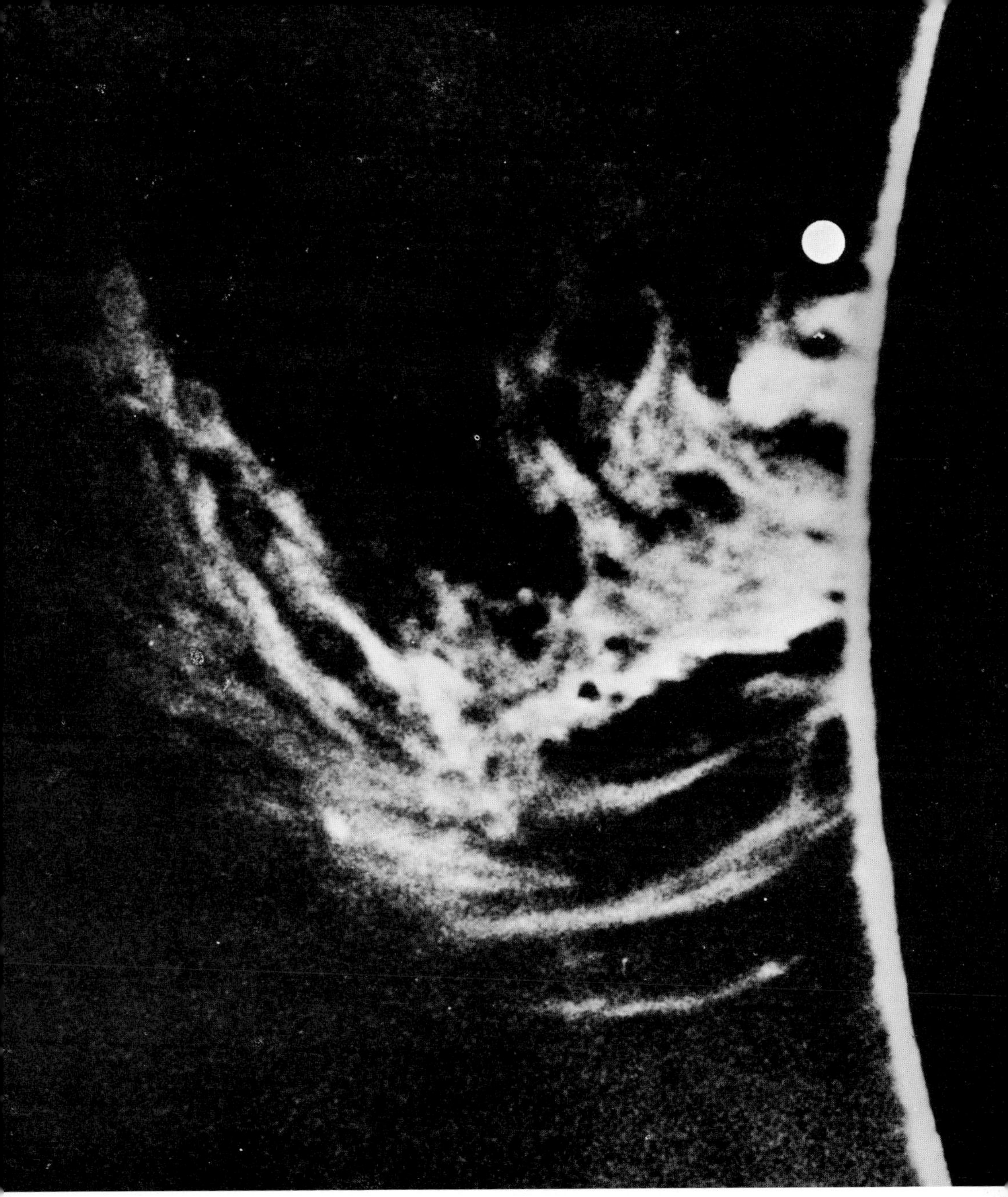

Flames leaping 140,000 miles from its rim graphically demonstrate the prodigious heat of the Sun, a constantly burning hydrogen explosion

An alert photographer caught Russia's Sputnik I as it streaked across the Canadian sky in 1957. Use of a time exposure permitted him to photograph the trail of the satellite

An attendant adjusts an antenna on a replica of Sputnik I, the first piece of solid matter to escape Earth and its atmosphere

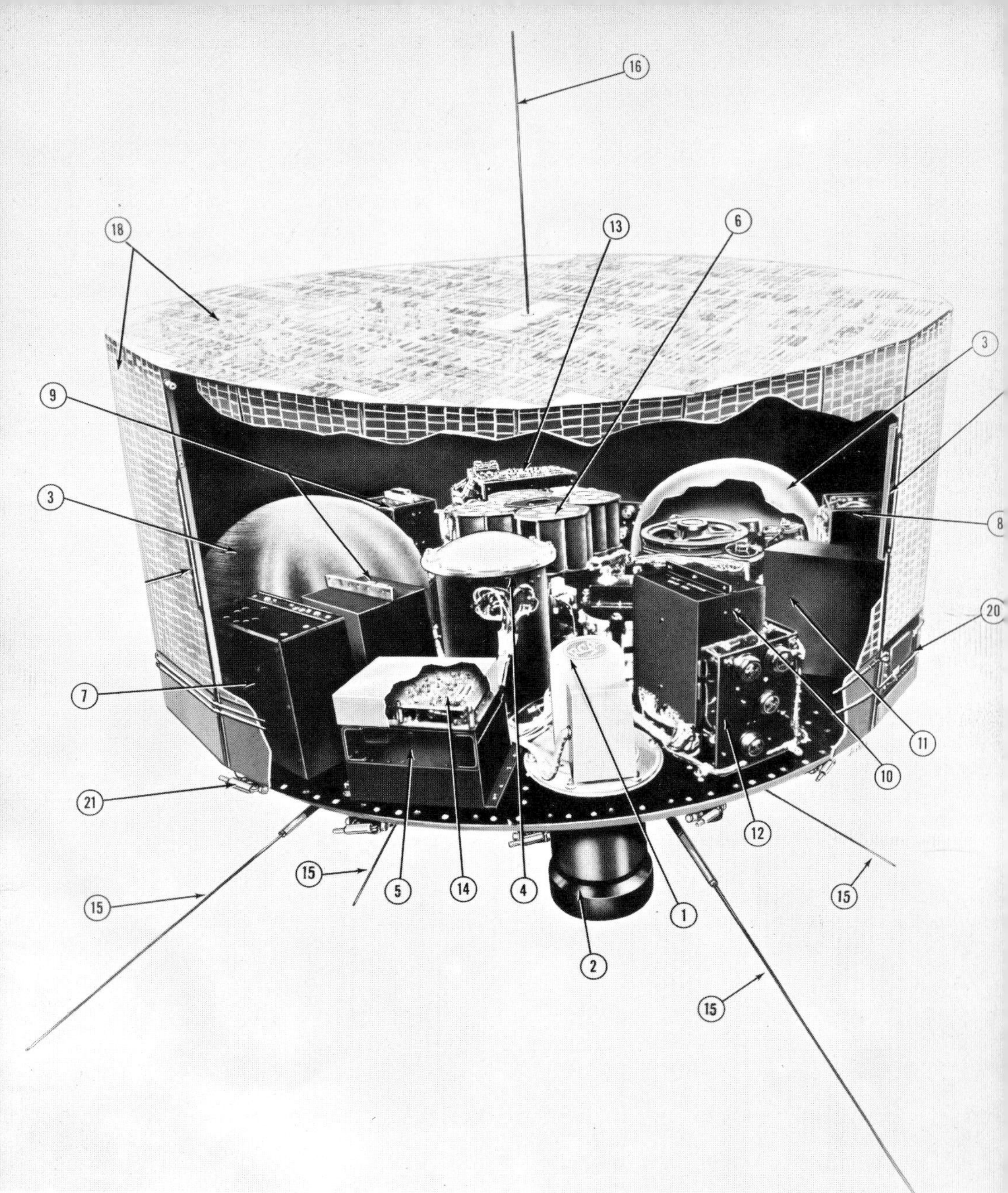

The U. S. Tiros weather satellite is crammed with equipment, including 1) TV camera, 3) tape recorders, 5) TV transmitters, 18) solar cells among other bits of mechanism most of which supply or regulate power

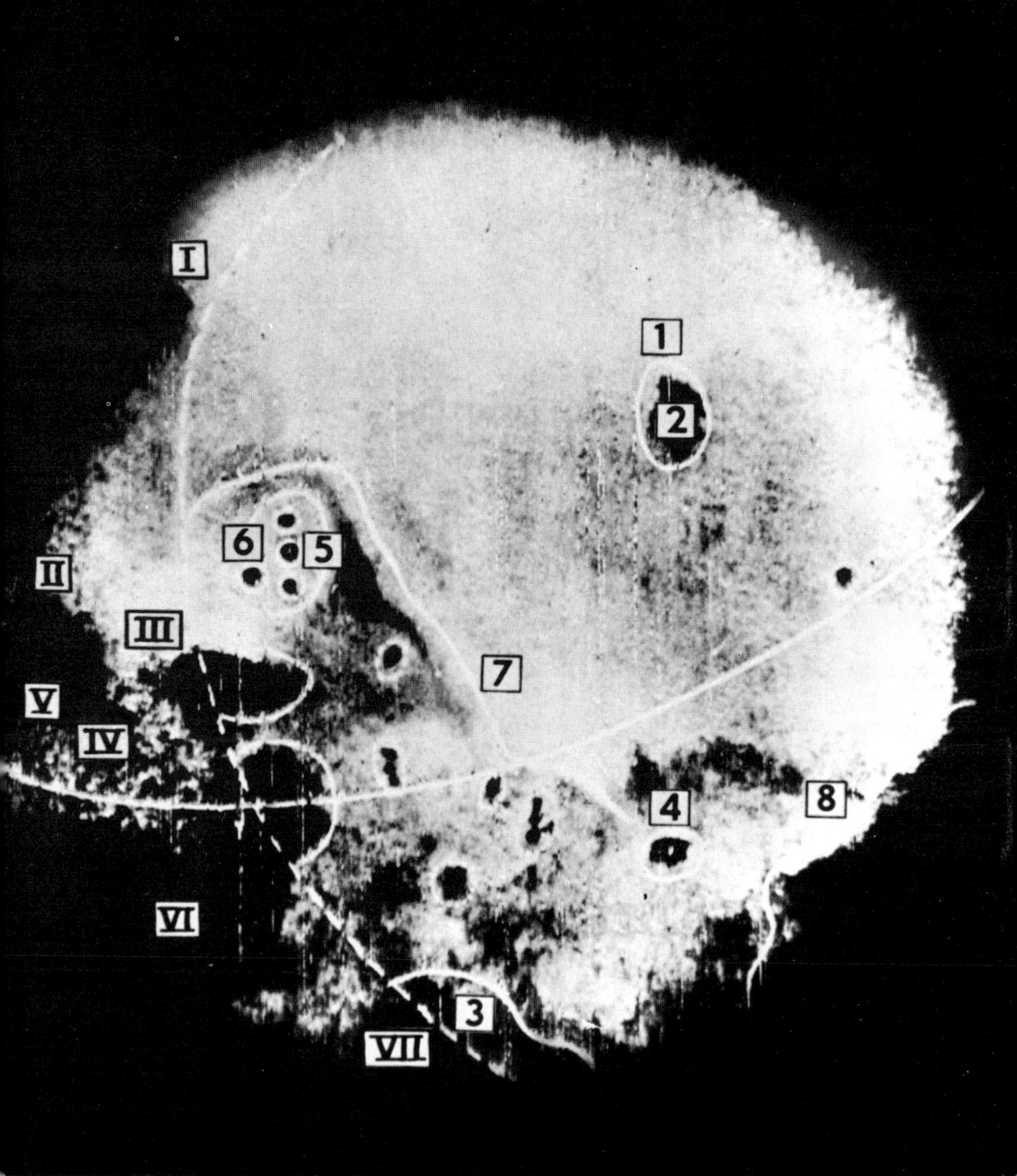

The back side of the Moon as photographed by a Moon satellite. The numbers refer to identifiable features of landscape to which are given such names as 1) Moscow Sea, 2) Astronauts Bay, and 8) Dream Sea

A "moonwatch" team tracks the path of a satellite passing near Fort Worth, Texas. Such teams give assistance to organizations observing celestial phenomena and also provide much enjoyment for the participants

Archimedes was, let us make clear, the maverick of ancient mathematics. He craved practical applications of mathematical principles, and to this day he is more famous for his mechanical inventions than for his important contributions to pure mathematics. A short time after his arrival in Egypt he invented the so-called water screw to lighten the awful labor of man and beast in pumping the waters of the river Nile to irrigate the fields. The device consisted essentially of a large wooden screw fitted into a long wooden cylinder. The bottom end of this cylinder was dipped into the Nile and the shaft was turned to lift water to the upper end of the cylinder, from which a constant stream flowed. This water screw worked in much the same way as the screw in a meat grinder pushes out the meat. The screw was turned by a water wheel, which was itself turned by the current of the river. Thus the stately Nile itself was compelled to do the work of delivering water into the countless irrigation ditches of Egypt. And, what is more, this water pump of Archimedes, invented over 2,000 years ago, is still used by the fellaheen – the peasants – in Egypt today.

Very likely these extracurricular mechanical ventures of Archimedes did not sit too well with the rest of the faculty. But they must have been shaken indeed when Archimedes presented his solution for one of the most baffling problems of mathematics of his time – namely, how to calculate the volume of bodies involving roundness, such as cones, cylinders, and spheres. Archimedes had suspected that the proportions of the volumes of these three bodies, if they are regular and their bases and heights are equal, must be as one, two, and three, and he had successfully tested the idea by means of wooden models made for him by a local carpenter. The models consisted of three cones, one hemisphere, and one cylinder, each with the same circular base and same height.

Armed with these "props," Archimedes arranged for a lecture

before the faculty, hinting at the announcement of a sensational mathematical discovery.

When he read the title of his paper, "On the Volumina of Round Bodies," the faculty became tense. This was a problem every one of them had tackled in vain for years. Could the brash young man have the solution?

Archimedes first announced his result, writing down the simple proportions:

$$\begin{aligned} \text{Cone} &= 1 \\ \text{Hemisphere} &= 2 \\ \text{Cylinder} &= 3 \end{aligned}$$

A hush fell over the audience. Everybody leaned forward to learn the proof of this sensational statement. Yet instead of a lengthy mathematical deduction, Archimedes produced his wooden props and a pair of scales. First he weighed the three cones against the cylinder – and they balanced. Then he replaced two of the cones with the hemisphere – and the scales remained balanced. Lastly, he weighed two of his cones against the hemisphere – and again they balanced.

Instead of applause, Archimedes drew icy silence.

Finally, a fourteen-year-old boy rose. He was Apollonius of Perga, a famous mathematician and member of the faculty while still in his teens. He had already won renown for his study of a family of curves of higher mathematics: the conic sections. He had described their strange properties and had given them their names: ellipse, parabola, and hyperbola. These amazing accomplishments had won him his place on the faculty while still a boy. Yet this is what Apollonius said:

"Mr. President, Gentlemen of the Faculty! I move that Archimedes be permanently banned from the university of Alexandria, because he has contaminated the pure spirit of mathematics with dirty matter!"

The position of Archimedes in Alexandria became untenable, because he had committed the worst crime against the spirit of mathematics. Mathematical proof can never be rendered by making experiments – only through pure logical argument.

Archimedes returned to his home in Syracuse. There he continued his creative career, discovering the principle of the lever and inventing the pulley, and formulating the principle of flotation that is named after him. In his later years he even solved theoretically the very problem that had caused his downfall in Alexandria: he was the first to compute the value and to describe the strange nature of the number π, "pi," which is the ratio of the perimeter of a circle to its diameter. Now it was possible to compute mathematically the volumes of round bodies without recourse to a pair of scales.

But Archimedes never lost his taste for practical inventions. The first eminent scientist in history to be called upon to serve his country in war, he invented a most efficient catapult which showered rocks on the Roman soldiers who attacked Syracuse under the general Marcellus. He constructed a levered grappling hook with which he lifted Roman ships and dropped them again, shattering them. He set up a large parabolic mirror which, concentrating the rays of the Sun, ignited the ships' sails.

With his flair for practical inventions Archimedes shows a strong kinship with our modern thinking. Even his mathematical papers read as though written in our time. Yet it would be wrong to see in Archimedes merely a highly gifted inventor. In his heart he was a Greek, faithful to their mathematical spirit, and he was actually much in sympathy with those who ousted him from the University of Alexandria. He never looked upon his ingenious hardware as something worthy of the true spirit of science. Rather, he called it "geometry at play."

The Roman biographer Plutarch, who lived about 1,900 years ago, wrote about the admirable scientific attitude of Archimedes:

> He possessed so lofty a spirit, so profound a soul, and such a wealth of scientific knowledge that, although these inventions had won for him the renown of more than human sagacity, yet he would not consent to leave behind him any written work on such subjects; rather, regarding as sordid and ignoble the business of mechanics and every sort of art which is directed toward practical utility, he placed his whole affection and ambition in those purer speculations where there is no admixture of the vulgar needs of life; studies, the superiority of which to all other is unquestioned, and in which the only doubt can be whether it is the beauty and grandeur of the subject examined, or the precision and means of proof that most deserve our admiration.

Archimedes met with a tragic death. All his ingenious weapons were unable to prevent the final fall of his home town against the superior numbers of the Romans. Marcellus had given orders to take the old wizard alive, but Archimedes was found by a Roman soldier who did not recognize him. Busy with a geometrical problem that he had drawn in the sand of his courtyard, Archimedes shouted at the soldier walking across his drawing: "Don't disturb my circles!" The enraged soldier lifted his sword; Archimedes asked, simply, for time to finish his mathematical proof; but down came the sword.

Upon his tombstone Archimedes himself had already carved designs representing his greatest accomplishment; a sphere and a cylinder, whose volumes he had been the first to compute. In the end he was a greater mathematician than any of them.

Much though the lofty thinking of such men deserves our admiration, their approach to scientific inquiry had severe limitations. Having developed the foundation for science — namely, mathematics — by sheer force of mind — they made little further progress. Experiment is essential to the understanding and, hence, the mastery of nature. The laws of nature can never be unraveled through abstract thinking alone; they must be found out and finally proved by experimentation.

Fifteen centuries after the great classical philosophers, during the Renaissance, there appeared a man of equal ability but with a different approach: the famous Italian scientist Galileo Galilei, credited more than any other in his time for introducing the experimental method. This is not to say that no one else had experimented. The English mystic and scientist Roger Bacon in the thirteenth century had conducted experimental studies that were both ingenious and varied. The alchemists, culminating in the great Paracelsus, had been experimenters. The sculptors and painters of the restless Renaissance age, with their unprecedented feel for the reality, the this-worldliness of nature, experimented not only with their chisels, paints, and brushes, but with the laws of perspective. These permit an artist to create the impression of depth, the illusion of three dimensions, on a flat piece of canvas. Even mechanical sighting devices were used in order to "deform" the art object in such a way that it would create the illusion of depth. The famous German painter Albrecht Dürer once created a number of interesting woodcuts in which he gave away such a trick of his trade. He shows a sighting frame used by the artist to "aim" the various points of the object to be drawn; to finish the drawing the frame could be swung out of the way.

The new drive toward experimental science as represented by Galileo, then, was only one aspect of a new viewpoint being taken by most creative people of the time. It appeared in sculpture and in painting; and in music and architecture and, in fact, in philosophy too. The human understanding had come down to earth. Here was the beginning of modern science.

As we look back, the progress seems immense. What genius has the experimental method set free! But it would not all have been possible without the spirit of the founders of mathematics, the first to demonstrate the powers of pure reason.

3 • *The Blue Planet*

The measurement of the size of Earth by Eratosthenes was perhaps the most magnificent example of the space-conquering power of the mathematical genius of the ancient Greeks. Up to the times of Alexandria, man's interest in his world was purely geographical — in fact, Eratosthenes himself was a famous geographer. But his feat was not just an accomplishment in the field of geography. Here was man, for the first time, seeing Earth as a whole — a celestial body. Eratosthenes put Earth among the stars.

In times after, many other important turning points in the intellectual history of man were also linked to revolutions in thinking about the nature of our planet. The idea that Earth is a sphere, not a disk, sparked the great geographic explorations of the fifteenth and sixteenth centuries — that magnificent age which brought all the continents within the ken of Western man. The recognition that Earth is a planet and revolves around the Sun created modern astronomy. Today our planet is still far from completely explored, and every year the "Earth sciences" are assuming more and more importance. Only recently scientists arranged the greatest single international research effort, the International Geophysical Year, to give Earth a thorough physical check-up and to learn more about the physical forces of the solar system and universe of which our planet is a part.

Since we live on the surface of Earth, it is difficult to step back and get an unbiased view of it as a whole. Perhaps our impressions of Earth would be much different and more representative of its true nature if it were not our home planet. As we now attempt a personality study of this rocky ball hurtling through space, let us be fanciful for a moment and try to see our home as others might see it.

You are a member of a technically superior race that lives on a planet which revolves around a star somewhere in the vast depth of the galaxy. Your people long ago solved the problems of interstellar and intergalactic flight. Of course, the distances between the stars are much too great for any kind of old-fashioned space travel such as crawling along at the speed of light. That way would take something like thirty thousand years to make the trip from the center of our galaxy to the outer regions of its sprawling spiral arms, where our Sun is located. You would not live long enough to see even the first tenth of the trip, even though your people have found ways to keep a man hale and healthy much longer than on Earth.

Forget travel at the speed of light; let's say you have a way of simply transferring yourself from one place in the Milky Way galaxy to another. One day you go out with an expedition to explore the outer fringes of the galaxy. By the sheerest chance, along the way, you enter the vicinity of our Sun. This not particularly bright star was utterly invisible to the naked eye on your planet. From there it had been detected only on a photograph taken with a very large telescope, where it appeared as a faint microscopic point among thousands of others. This quite ordinary star, of a type common in the population of our galaxy, shows nothing to arouse special curiosity. But your course is taking you within a mere ten billion miles of it, and when you are that close to even an ordinary star you might as well go off course just slightly to take a closer look.

You are about three times farther away from the Sun than Pluto, the outermost planet of this system. The Sun is so far away that it does not show a disk to the naked eye, but is a mere point. It has a fierce light nevertheless, being almost a thousand times brighter than the planet Venus, the brightest light in the sky for Earth except for the Moon and the Sun itself. Toward the Sun, now, you turn, shifting to the old-fashioned rocket drive. At a speed somewhat less than that of a ray of light, you can expect to reach the Sun in something like fifteen hours.

Presently you are inside the outer orbits of the solar system. As the planets become visible one by one, you study them in detail with the sophisticated instrumentation of your advanced civilization. Earth men took centuries to determine the known facts about the nine planets, but you do it in a few hours. You find the planets, like the Sun, to be mostly common types that you have seen before on previous interstellar expeditions. All the planets are fairly ordinary, except for two.

There is planet Number Six – the one Earth men call Saturn. It is adorned with a system of giant rings that glitter in the sunlight like pearl necklaces. Similar rings you have seen around planets in other systems, but none so beautiful as these. The rings are of extreme thinness, thinner in proportion than a piece of tissue paper several feet wide. Their mathematical precision is superb – perfect concentric circular rings, as if drawn on the black curtain of the sky with a huge pair of compasses. Never

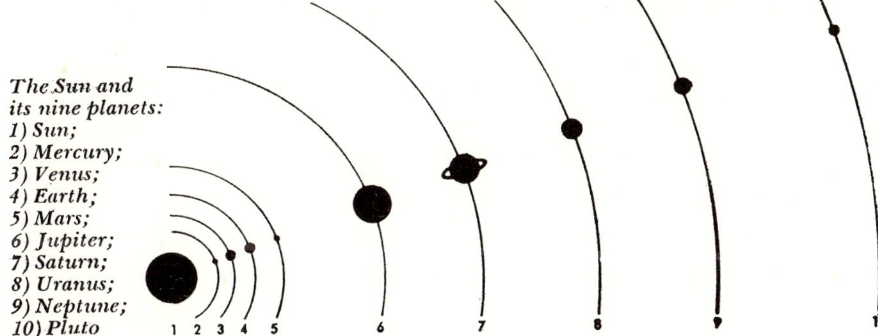

The Sun and its nine planets: 1) Sun; 2) Mercury; 3) Venus; 4) Earth; 5) Mars; 6) Jupiter; 7) Saturn; 8) Uranus; 9) Neptune; 10) Pluto

before have you encountered such a splendid ornament on a planet.

Then there is planet Number Three. This has, considering its size, a very large satellite; actually, the two bodies form what could be called a twin planet. Planet Number Three spins around its axis once every twenty-four hours, and wheels around its star in a little more than 365 days. But what sets this planet apart from all others — even when seen from a great distance — is its strange color. All the other planets are reddish, yellow, or white, only the giant outer ones having a tinge of green. But planet Number Three is blue — in fact, it is the only blue planet in the whole solar system.

As you go by, the full lighted face of the planet is seen glowing in soft colors. Subdued hues of orange and green are spotted with wide patches of dazzling white, but the blue is still there. The whole planet is suspended in a veil of bluish haze, and broad areas of its visible surface show a deep inky blue.

But there is something else that makes planet Number Three something special among all others. The Sun mirrors itself on the surface in a dazzling highlight that lends the majestic sphere a sweeping roundness. It appears to float in space like a glistening Christmas tree ball. This highlight is caused by a remarkable feature: most of the planet's surface is wet! Not a single other planet in the solar system has a liquid surface.

Now that you have a notion of how Earth, as a planet, looks to an other-worldly traveler from outer space, try taking a closer look at it with a still unbiased, fresh point of view. This is not easy, because we live on this planet and our notions of its features are distorted by their closeness. Because of the tremendous size of our planet we have, for example, consistently wrong concepts about the relative dimensions of everything connected with it. To get things into a perspective we can more easily comprehend, let us make for ourselves a model Earth measuring something like two feet in diameter — about twice as big as an ordinary globe. The scale is about one to twenty million. But our model is not an ordinary globe. Everything on and in the model is in exactly the right proportions. Now we are going to make some startling discoveries.

First check the weight of our Earth: here is where we get our first surprise. The weight is 1,300 pounds, as much as eight strong, grown-up men! Our planet is amazingly heavy, in fact; for its size it is the heaviest of all planets, because no other is made up of denser material. A large part of Earth consists of heavy metals; the center, probably, is a ball of iron and nickel, encased in a shell of rock.

While the weight of Earth becomes truly spectacular, other features shrink to insignificance. Take, for example, the atmosphere. We think of it as a tremendously deep ocean of gases, reaching far out into space. Actually it is but a very thin "haze" close to the surface. We have considerable difficulty in finding material to represent the extreme thinness of our atmosphere true to the scale of our model. Only the finest, most nearly invisible covering of Kashmir silk could begin to suggest the thickness of the atmosphere of our model Earth at its surface. A little smoke ring from a cigarette, spread evenly around the model, would amply represent the bulk of our atmosphere.

The oceans, together, occupy more than two thirds of Earth's surface. On our model globe they are true to scale not only in size but also in the amount of water they contain. If you touch our model in the middle of the Pacific Ocean, your fingertips become barely damp. Can "damp" be the right expression when we are describing the endless masses of water of the Pacific? It is the right word for our model, because here our oceans are less than a hundredth of an inch deep. If we collected all the waters from the seven seas on the model there would be about enough to half-fill a fruit-juice glass!

Besides sea water there is also fresh water on Earth. How much fresh water would we need for our model? One drop from an eyedropper would be too much; so we use an atomizer, which produces a droplet about one tenth as big as a drop from an eyedropper. The water in the droplet fills all the fresh inland seas of the Earth — the Great Lakes, Lake Victoria, the Aral Sea; it fills all the great rivers — the Amazon, the Mississippi, the Nile; it fills all the smaller lakes and rivers and the underground reservoirs of fresh water, and it supplies enough moisture for the rainstorms, too!

Not all of the water on Earth is in liquid form; much is in the form of ice. In fact, there is more ice on Earth than there is fresh water. Glaciers fill the broad valleys of high mountains in most regions. But the bulk of the ice is in huge ice sheets over or near the poles. To airline passengers on polar flights, the masses of the earth's ice must appear immeasurably great — the more so when they hear that the ice sheets in Greenland and Antarctica are, in spots, nearly two miles thick. Yet for Earth as a whole the total amount of ice is a mere frosting at the poles. The tip of an ordinary icicle that would melt almost instantly between your fingers: this much ice will be enough for our model. Spread out evenly over the polar areas, it will form a layer about as thick as a piece of thin cellophane.

On our model the highest mountains, such as Everest, and the deepest ocean trenches, such as the great 7-mile-deep cleft near the Mariana Islands, are hardly visible. The sphere looks almost perfectly smooth. This surprises us, because the usual relief models of mountain ranges and ocean basins are all grossly exaggerated: they *have* to be to make them useful at all!

With our model we do a trick to make the mountains and ocean troughs visible. We project the rim of the model, much enlarged, upon a screen, and compare the rim to a thin, perfectly circular arc drawn with a compass. Now we perceive the slight irregularities along Earth's rim: these are the loftiest mountain ranges and the deep ocean basins. The average depth of the oceans on our globe is not more than the thickness of two sheets of average writing paper, and the mountains are smaller than the wrinkles in a crumbled piece of cellophane that has been spread out again on a flat table.

By the same token, the solid stone mantle of Earth is extremely thin, too. Let's pull the United States out of the model in one piece — all of the United States that is solid. We get a nicely hollowed, rough-edged bowl thinner than a piece of eggshell. It is about 9 inches long, 6 inches wide, less than 1/10 inch thick. Very different indeed, from what we are used to seeing on flat maps!

Models are extremely useful for setting right many misconceptions about Earth and the dimensions of its features. One such misconception is that Texas is bigger than California. The distressing fact is that California is bigger than Texas. Texas is a rather flat state, with a low average elevation above sea level, while California is quite mountainous. Reckoned in terms of volume above sea level, California has more to offer than Texas. That there is more to California than to Texas can be shown, in fact, by weighing the two, if you ever have the chance.

But wait — a Texan can still insist that his beloved state is bigger than all the United States — including California, Alaska, Hawaii, *and even Texas itself!* Texas extends so far and wide over the globe that it has a slight but clearly visible curvature. Taking Texas out of our model earth and setting it upside down on a table, we find it shaped like a flat, slightly hollow pan. Now we scrape down to sea level the entire United States, including Alaska (which recently and rudely elbowed Texas out of first place) by shaving off all the mountains, hills, and plains, all the soil and rock, that lie above the level of the sea.

The shavings form a small heap which we pour into the hollow pan of Texas. But the pan is little more than half full! So the entire United States, including Texas, is smaller than Texas, which holds easily every bit of the fifty states that is above water. There is even a handle on the pan to carry them away.

The removal of a large section of the North American continent deprived our model of a portion of the solid crust. The red, molten interior is now exposed. We see how unbelievably thin the solid crust actually is: twice as thin, in proportion, as the shell of an egg. Almost 99 per cent of the entire mass of Earth is a molten or plastic; only a little more than one per cent is solid. However, it is a strange kind of liquid, if we can call it that, which fills our planet. When an earthquake strikes, these innards of Earth behave as though made of high-grade steel. The shock waves travel large distances at very great speeds, and this is possible only because the innards are very dense. In an earthquake, in fact, our planet behaves like a steel ball that is struck with a gigantic hammer.

It seems inconsistent that in an earthquake Earth should behave like a sphere of steel, but on other occasions behaves like a ball of jelly — runny jelly, at that. Even its hard, stony crust yields easily to the forces of deformation. Fortunately for us, Earth rotates rather slowly, so that it remains almost a true sphere. (Our model Earth measures only one tenth of an inch more across the equator than from pole to pole.) However, if we could cause Earth to rotate seventeen times faster than it does, centrifugal force at the equator would offset the force of gravity there. Then the planet would begin coming apart, with sections of ocean and land flying off like the sparks from a pinwheel.

Earth can be both as hard as steel and as soft as jelly at the same time only because it is so big. Things become relatively less and less rigid as they get bigger. Take a tiny ladybug, for example. It weighs less than one one-hundredth of an ounce. Yet it is so strong that it could easily carry a hundred of its sisters on its back. Only a Tarzan among men could carry three or four of his own kind. Man is much larger than a ladybug and so, in proportion, his structural rigidity is much smaller. Man, in turn, is smaller than a whale. These big animals are so weak that they can survive only with support of water — that is, when floating. A whale accidentally washed ashore suffocates as the weight of his great bulk collapses his lungs. All big animals are proportionally much heavier and thus clumsier than small ones. If the skeleton of a giant saurian and that of a small lizard were reduced to the *same* size, the bones of the lizard would appear far more delicate than those of the saurian.

The same is true for the structures of plants. It has been said that a stalk of wheat is a pure miracle of structural stability — far superior to any designs of man. If a wheat stalk were as tall as a giant smoke stack, with all its parts in scale, it would be only about one foot in diameter, yet would be carrying a weight of several tons of grain on top. This is often quoted as proof of nature's superiority to man in techniques of design. Actually, the example proves little. Man can easily duplicate the design of a wheat stalk over six feet high by using a stiff steel wire; this could easily carry the weight of a ripe ear of wheat on its top without breaking in the wind. By the same token, a stalk of wheat enlarged to the size of a tall smokestack would collapse at once; it could not stand up under its own weight. Even "nature" could not circumvent the facts.

Flimsiness increases with size. Skyscrapers sway noticeably even in moderately gusty weather. The sway and vibrations of great suspension bridges like the Golden Gate Bridge can be felt by everyone who crosses on foot. Proportionally, such a giant bridge is weaker than a plank thrown across a brook.

Earth's jellylike softness is the reason why the planet is an almost perfect sphere. A soft body of such size could not possibly have any corners or edges, as a cube does. Protruding parts would immediately collapse under their own tremendous weight. In other words, gravity would pull all the protruding masses together with the rest into spherical shape. Earth is a big ball, a nearly round one, because it is in this shape that all the masses that form the huge bulk are crowded about as close together as possible.

Our Earth is much more delicate than a Christmas tree ball. It could not even stand up under its own weight. Suppose we could lay our planet down on the surface of a much larger planet where all things weigh as much as here on Earth. Setting down Earth as tenderly and carefully as possible, we still cannot prevent its total destruction. As Earth touches the ground, it flattens out at the point of contact like a drop of honey touching a saucer. Huge cracks appear all over its surface and break wide open, allowing molten masses to pour forth. Steam shoots up as the oceans evaporate. And soon old Earth has collapsed and sunk to a shapeless, sticky, hissing mass.

The pull of the Sun, which holds Earth in its orbit, is gentle and constant. Earth's rotation is easy, leisurely. The planet floats along without resistance or interference, and without support. May its weakness never be put to a test!

4 • *The Clock Stops at the Moment of Death*

Our own Earth is by far the most interesting planet of all. This is said not out of narrow-minded pride in our own cosmic "home town" – there are many potent scientific facts to prove it. For today we can safely conclude that among the nine planets in our solar system only Earth harbors life in all its colorful variety. Life is a powerful force that has a tremendous influence upon the physics and chemistry of the planet's crust, its oceans, and its atmosphere. This is what makes the Earth sciences so interesting: almost every one of its problems involves many sciences at the same time, and these problems more often than not relate intimately to human affairs. A single science story will prove our point: a story that could not have happened on any other planet in our Sun's family.

We begin with a tiny atomic particle hurtling through the endless spaces of our galaxy. Like a super-fast tiny bullet it races forward with a speed almost as great as that of light – 186,000 miles every second. But the expanse of space is so unbelievably vast that it takes the little bullet thousands of years to make any real headway across the galaxy. Its flight is aimed, by chance, at the needle-sharp speck of a star. The star is a blazing sun – our own Sun – but its floods of light are dimmed to a mere point by the fearful distance.

The atomic bullet hurtles on and on, and in the course of hundreds of years the Sun appears stronger and brighter. During the last few days and hours of the bullet's flight, the Sun becomes a blazing disk, larger and larger. The bullet crosses the orbits of the outer planets of the solar system, passes Jupiter and Mars, and heads squarely at the Earth.

The disk of Earth grows from the size of the full Moon to a huge wheel that covers half the heavens. With a thousand miles to go, the bullet plunges into the outermost layers of the Earth's atmosphere, races through them with hardly slackened speed, and finally, some 100,000 feet above the ground, crashes into an atom of Earth's air. The impact is so terrific that both bullet and target are smashed into pieces that fly off in all directions. One of these fragments crashes into a nitrogen atom drifting nearby, and as the fragment hits, it changes the chemical "personality" of the nitrogen atom completely.

Before getting on with our story, we must see how a fragment from one atom can change the personality of another. There are essentially three kinds of building blocks in an atom. One is an extremely tiny, light-weight particle that carries a negative electrical charge; it is an electron. The other two building blocks are equally small but almost 2,000 times heavier than the electron; their names are "proton" and "neutron." The proton bears a positive electrical charge of the same strength as the electron, so that one proton and one electron can neutralize each other completely in their charges. Within the structure of the atom, the protons and the neutrons are fused together into a dense heavy ball – the atomic nucleus – which sits in the center of the atom. The electrons swarm around the nucleus like a cloud of tiny flies; or like planets around a sun. The electrons are held in place not by gravity but by the electrical attraction that acts between them and the protons in the nucleus.

The chemical nature of an atom is determined by the number

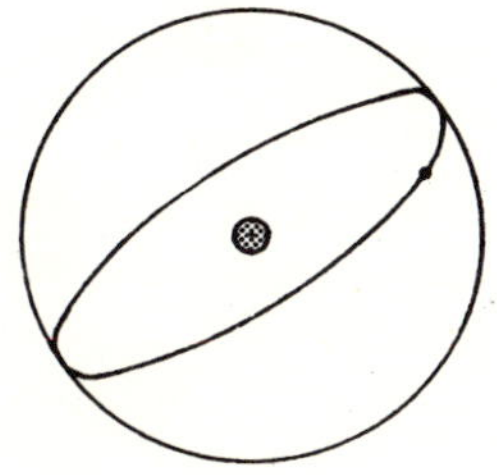

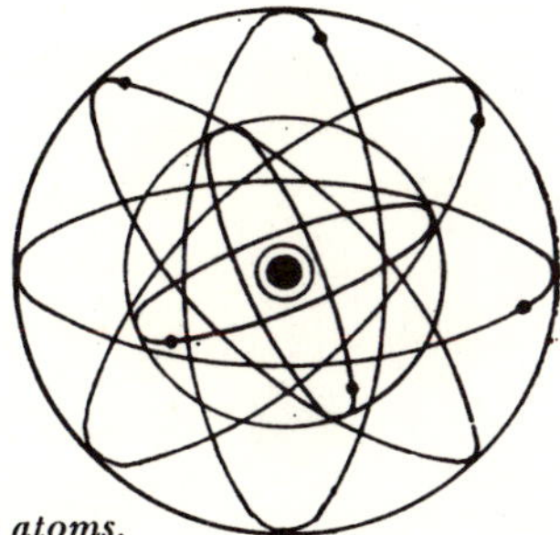

Schematic diagrams showing Hydrogen and Carbon atoms.

of protons it has in its nucleus. Hydrogen is the simplest of all atoms: it has a single proton as its nucleus. An atom of helium has two protons in its nucleus, carbon has six, calcium has twenty, gold has seventy-nine, and so on. The atoms of each chemical element can simply be called off by a number. There are neutrons in the atomic nuclei, too, but their number is not important to the chemical nature of an atom; the neutrons only add to the weight of the atom.

Now, back to our story. What happens when the atomic intruder from outer space crashes into the nucleus of an atmospheric atom?

The impact smashes both the bullet and the target nucleus to pieces, and their component protons and neutrons fly apart in all directions. One of the fragments, a neutron, hits the nucleus of our nitrogen atom and sticks there.

Before the collision happened, the nucleus of the nitrogen atom had seven protons and seven neutrons. The seven protons made it a nitrogen atom; the atomic number of the chemical element nitrogen is seven. Now the nitrogen nucleus has eight neutrons — but not for long. The nucleus of a nitrogen atom cannot hold an extra neutron; so, something like a hundred-millionth of a second later, our nitrogen atom throws out a particle to recover its balance. But the particle thrown out is not the neutron that came in; it is a proton. And this is where

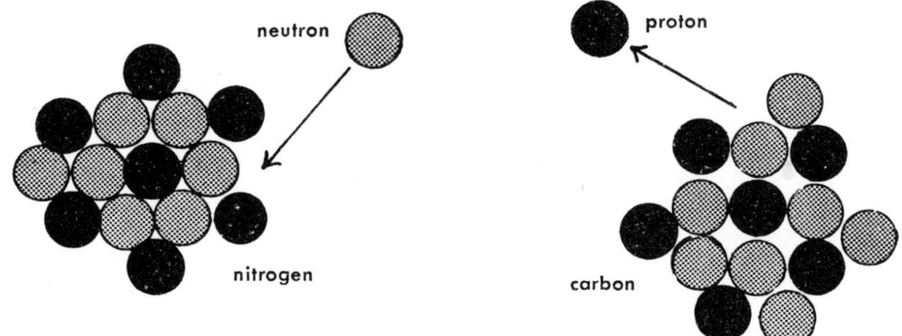

A neutron enters the nucleus of a nitrogen atom, making it unstable. A proton is emitted to balance it again and an atom of Carbon-14 results.

our nitrogen atom suffers a change in its personality. Before, it had seven protons and seven neutrons in its nucleus, and now — a split second after these dramatic events — it has six protons and eight neutrons! That makes it an atom of carbon, because the atoms of this element have six protons in their nuclei.

This new carbon atom, however, is a little different from the ordinary carbon atoms that abound in such materials as coal, graphite, and diamond. The normal carbon atoms have only six neutrons, while ours ends up with no less than eight! Whereas ordinary carbon atoms have twelve units of weight (protons and neutrons weigh practically the same), our special atom has fourteen. Hence it is called "carbon-14."

But this is not all. The nuclei of carbon-14 atoms are top-heavy. They are unstable and can survive only a certain length of time. Then something "snaps" inside the overburdened nucleus, and it again changes its personality. The thing that snaps is one of the eight neutrons in the nucleus.

A neutron can be pictured as a tight package composed of a proton and an electron. Since a proton and an electron have opposite electrical charges of equal strength, they neutralize each other perfectly and, when combined, form a tiny neutral particle — a neutron. So when finally a neutron inside the carbon-14 atom snaps, the electron is thrown out of the nucleus with great force, and the disintegrating neutron is transformed into a proton. After the electron has been ejected, the nucleus is left with only seven neutrons, and the extra proton formed by the splitting neutron is added to the other six. This makes seven protons in all, and the atom has thus been re-transformed into a nitrogen atom not different from the one it was before. This is indeed a remarkable cycle.

How long does it take for any individual carbon-14 atom to snap one of its neutrons and reform itself into a nitrogen atom? Nobody knows. It could happen in the next second, after a week, or only after thirty thousand years.

Yet, no matter when this happens, another carbon-14 atom is likely to be created somewhere in the atmosphere of Earth. For collisions like the one that gave birth to our atom are taking place in the atmosphere all over the world all the time. There are billions of cosmic bullets crashing into the atmosphere in every second. These bullets are the particles of cosmic radiation that traverse outer space in all directions with their incredible speed. In every second, billions of atomic explosions occur as these tiny projectiles smash countless air atoms to smithereens; there are countless neutron fragments flying about that crash into countless nitrogen atoms nearby causing them to change into carbon-14 atoms.

The creation of these atoms in the air has been going on for millions of years. Each of them becomes a nitrogen atom again when its unpredictable hour strikes. In the course of time, an exact balance has been established between the number of births and re-transformations of carbon-14 atoms, so that their worldwide number stays exactly the same at all times. Of all the carbon atoms that are in circulation all over the world, one out of about one trillion is a carbon-14 atom.

Assume, now, that one carbon-14 atom enjoys a long, long life. After its birth it does not stay single for long, but soon picks up a lonely oxygen atom and then another. With these two companions it forms a compact package held together by the forces of chemical bond: the three atoms form what a chemist calls a carbon-dioxide molecule.

This molecule drifts down from the 100,000-foot level where it was born. When it reaches a height of 45,000 feet, it is caught in a violent jet stream — a current of air racing around the globe along a winding path, west to east, with a speed of hundreds of miles per hour. After a wild journey of an hour or so, our molecule drops below the stratosphere and is caught on a tiny ice crystal of a cirrus cloud. As the cloud sinks lower, the ice crystal grows in size and begins to fall slowly. Reaching warmer layers of air, it melts, and as a raindrop it falls into the ocean.

For long years the molecule, tossed about in the waves, following the currents, is a sea voyager. Then one day it is sucked up by a tiny, one-celled sea plant. By complex chemistry, the plant takes hold of the carbon-dioxide molecule, tears off the two oxygen atoms, and uses the carbon atom as a building block for its growing body. In time the plant is eaten by a shrimp, the shrimp is eaten by a herring, and the herring is eaten by a seal. Each time, the carbon atom becomes part of the body of the animal. However, it stays in the body of the seal only a short time. The seal's metabolism causes the carbon atom to join up again with two oxygen atoms and thus form another carbon-

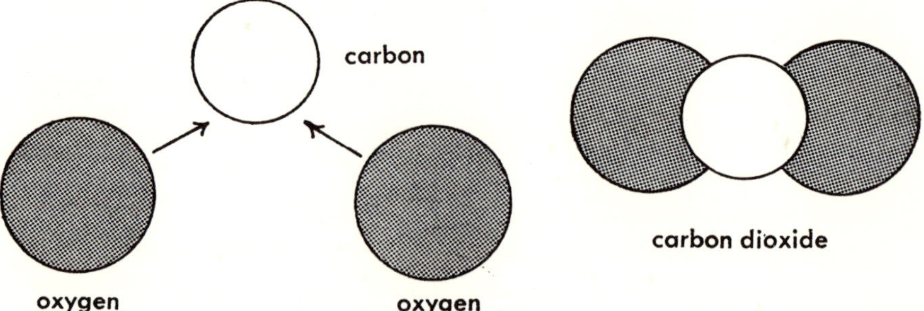

dioxide molecule. This molecule is picked up by the seal's bloodstream, enters the lungs, and is exhaled into the air.

The molecule drifts across land and sea, perhaps for centuries. At last it is caught by a young redwood tree, and as before it becomes a building block for plant tissue. Deposited in the trunk of the young tree, it lies buried there for centuries while the redwood sapling grows to gianthood. The tree lives on for thousands of years – until its hour strikes. A devastating forest fire rages through the redwood grove, and the old tree is consumed in the blaze. The trunk burns to ashes, our carbon atom again joins up with two oxygen atoms, and as part of yet another carbon-dioxide molecule it drifts away with the dense cloud of swirling smoke.

After traveling hundreds of miles, the molecule is picked up by a blade of grass on a prairie. In late summer of that year a buffalo comes along and eats the grass. The carbon atom, traveling through the body of the buffalo, arrives in the thighbone. The buffalo soon falls prey to a prehistoric Indian hunter, who drags the beast to his cave and cooks himself a hearty meal. When he is full, the Indian, whose table manners are not very refined, throws the charred thighbone over his shoulder, and it lands on a trash heap in the back of the cave.

A few years later the prehistoric hunter dies and he is buried in the cave. Eventually the cave roof falls in and covers the body of the hunter, together with the thighbone of the buffalo, beneath a thick layer of sand and rock.

Thousands of years pass. Then, in the year 1926, an archeologist digging near the town of Folsom, New Mexico, discovers the skeleton of the stone-age hunter. The age of the skeleton is estimated at 13,000 to 15,000 years, the estimate being based on a study of the rock strata there. For years thereafter, "Folsom man" is considered the oldest-known inhabitant of the North American continent.

The charred bone is dug up, too, and sent to a museum in the eastern United States. For about a quarter of a century our carbon atom lies in a glass box in the museum, still holding the secret of its strange birth and of its future.

Shortly after World War II, a nuclear physicist comes to the museum and requests a small piece of the charred bone. He takes it to his laboratory, grinds it up, and extracts all the carbon from it. Then he accurately weighs out a sample of the carbon and puts it into the counting chamber of a Geiger counter. Our carbon atom is one of the atoms put into the counter.

Three hours and twenty-one minutes after the beginning of the experiment, our carbon atom finally lets go. One of the neutrons in its nucleus disintegrates into a proton and an electron. The proton remains in the nucleus; and so, after something like 20,000 years, our atom becomes a nitrogen atom again. The ejected electron, however, lances through the tube of the Geiger counter and causes it to click. The automatic recording device hooked up with the counter advances by one notch.

After twenty-four hours, the physicist finishes his experiment. He takes his reading from the recording device and jots it down. He busies himself with his slide rule, then writes down the number "9,900."

What has the physicist been doing? He has measured the age of the charred bone, and he has found that the buffalo was killed some 9,900 years ago. This, then, must also be the age of Folsom man, because the bone was found with him. Thus Folsom man, after all, turns out to be several thousand years younger than previously assumed.

Our carbon atom has served its role as part of a clock that started to run at the moment of the buffalo's death.

Just how does the clock work?

Every living thing constantly uses carbon from Earth's atmosphere to build its tissues. Since the ratio of carbon-14 to carbon-12 in the atmosphere is always in the ratio of one to about one trillion, the ratio in living things is the same. However, when an animal or plant dies, it ceases to take in any more carbon from the atmosphere. At the moment of death, therefore, the ratio of carbon-14 to carbon-12 atoms begins to change. The carbon-14 atoms, one by one, break down because of their radioactivity, and they are not replaced by new ones from the atmosphere. The carbon-12 atoms, on the other hand, are stable, and remain the same. The result is a gradual decrease in the proportion of carbon-14 to carbon-12 in the dead tissues. Since the decrease occurs at a precise mathematical rate, the proportion that exists in the dead tissue at any time after death indicates just how long ago death occured.

The grass of the prairie, so long as it lived, kept the one-to-a-trillion ratio. The buffalo that ate the grass likewise maintained this ratio in its body. But as soon as the buffalo was dead, the ratio in its thighbone began to change. Just about 9,900 years later the physicist measured the change in the ratio, and that told him how long ago the buffalo had met his death.

What about the mathematics of this process?

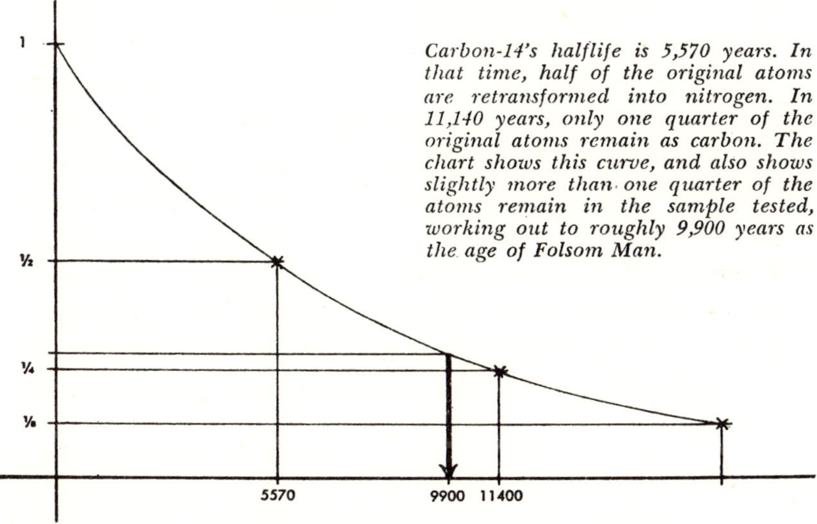

Carbon-14's halflife is 5,570 years. In that time, half of the original atoms are retransformed into nitrogen. In 11,140 years, only one quarter of the original atoms remain as carbon. The chart shows this curve, and also shows slightly more than one quarter of the atoms remain in the sample tested, working out to roughly 9,900 years as the age of Folsom Man.

It has already been explained that there is no predicting when any particular carbon-14 atom will let go. How then can a decrease in carbon-14 atoms in dead tissue indicate the date of death? The answer is that although the behavior of any particular atom is unpredictable, the behavior of a large group of them — millions or billions — is predictable. In roulette we can't tell what the *next* jump of the ball will be, but we can tell — if the roulette wheel is honest — how many times the ball would fall on each number if the wheel is turned billions of times. In the same way, the future behavior of large numbers of carbon-14 atoms can be predicted accurately.

We have found that half of a large number of carbon-14 atoms will be changed to nitrogen atoms after exactly 5,570 years. In the case of those that are unchanged, half of these will also let go in another period of the same length of time. After about 30,000 years, the amount of carbon-14 remaining is too small a proportion to make accurate dating possible.

Dr. Willard F. Libby, who developed this dating process, was given the 1959 Nobel Prize in chemistry. His method has been used to determine the age of many other organic materials, such as the wood in an ancient Egyptian tomb and the fibers in the dress of an Aztec princess.

And so we come back to the start of this chapter and our statement that Earth is the most interesting of the planets.

Our reason was that the existence of life on Earth is constantly affecting in some way all the events that are happening on Earth's surface every minute of the day, year after year, century after century. Today, in this time when the secrets of the atom are being unlocked, the story of the nitrogen atom reminds us that the whole of creation is a vast, constantly changing fabric of racing atoms, among which — on Earth, at least — the influence of living things is always present.

5 • *The Solar Powerhouse*

Not counting Earth, the Sun is the most important body of the solar system in terms of human affairs. Through its overpowering gravitation, it acts as a stabilizer to keep the delicate dynamics of the planetary motions in order, and with its life-giving radiation it prevents Earth from becoming an ice-encrusted desert.

A few years ago, an imaginative writer of science fiction finished off the world by depriving it of the warming rays of the Sun. In his story another star approached the solar system and, passing nearby, played havoc with the orbits of the planets. The star's gravity destroyed the delicate balance of the whole system. Earth was torn loose from the Sun's grip and set adrift in the eternal cold of space.

Deprived of the warming energies of the Sun, the oceans of Earth began to freeze. The planet was soon covered with a solid armor of ice, and the hollows were filled with steaming pools of liquid oxygen and nitrogen – the meager remainders of the atmosphere. Gradually these pools, too, froze solid. The once-mighty disk of the Sun shrank to the size of a pinhead, and cast a weak and dismal light over the desolate surface of Earth. In another century the Sun was a mere remote star among many others – just another cold point of light. Meanwhile man, exhausting his puny resources, fought a losing battle, and finally succumbed.

The dismal visions of that science fiction writer will probably never come true. Earth is firmly anchored in its orbit around the warming Sun. The spaces that stretch between the stars are so immense that the odds against a collision or a near miss between any two of them are almost immeasurably large.

But the story does show in a dramatic way how dependent all life on Earth is upon a steady supply of solar energy. In that story it was not the merciless encroachment of the cold that ended life. The peoples on the planet knew for years what their fate would be, and once the astronomers had calculated what would happen, the world's statesmen made extensive preparations for human survival by means of the internal heat of the globe. Burrowing a few hundred yards below the surface, the peoples of the world remained protected by the almost perfect insulation of the thick layers of rock above. Finally, however, they starved. Without sunshine the planet's vegetation could no longer carry on its food-producing activities. The eternal cycle of life was interrupted.

The radiant energy from the Sun is the source that powers all the processes of life. In fact, it is the ultimate source of nearly all power on Earth.

When we drive a car, we are using fossilized solar energy to propel it. The gasoline in the tank has been extracted from crude oil, formed beneath the surface of Earth over periods of millions of years. Ages ago, living organisms enjoyed the warming rays of the Sun that shone upon them and gave them a chance to live and to grow. As they died, in some areas their decaying bodies accumulated in great numbers, they were buried under sand and silt which turned to rock, and oil from the remains gradually filled the spaces between the rock grains.

When we turn the switch of our television set, it is solar energy that provides the power. The electric energy of our city supply lines comes from the power station where coal is converted into electric power. Coal, too, is fossilized solar energy. It is the remains of masses of plants that lived during past geological ages, used solar energy to build up their tissues, died, and were covered by gradually increasing masses of sand and mud. Pressure slowly carbonized them — turned them into the black, combustible coal we know.

Or, the power station may draw its energy from the waters accumulated behind a dam. These waters, too, represent solar energy. Every day in the year, the Sun evaporates about one thousand billion tons of water from the sea up into the atmosphere, and much of this monstrous mass of water soon rains down on the continents. Thousands upon thousands of tons of this water flow from streams into the reservoir behind the dam. Water coming through a spillway in the dam in a continuous stream is directed against the blades of huge turbines, and these drive generators that produce our electricity.

A farmer uses a windmill to pump underground water up into a high tank, from which he pipes the water down to his house and barn. The pump is driven by the windmill. The vanes of the windmill are pushed by a breeze. The breeze is caused by the movement of currents of air which are set up by unequal heating of the ground and air by the sunlight: warmed air expands and rises, while cooled air contracts and sinks. So, it is solar energy that finally drives the farmer's pump.

Sometimes great amounts of stored solar energy are released in concentrated, devastating form. Ultimately the Sun is responsible for tornadoes and hurricanes, floods, and forest fires. All too frequently do we suffer from the disastrous onslaught of the powers of nature. It is hard to visualize that the energy unleashed in these terrifying forces was once embodied in the friendly light of the Sun.

A tiny fraction of the total energies that operate across the face of Earth does not originate in the Sun. A rock slide that comes tumbling down a mountain slope derives its energy from the internal forces of the earth that once heaved the mountain up. The energy of earthquakes is also of terrestrial origin. Volcanoes, geysers, and hot springs have their source in the heat of the Earth's interior — heat due partly to radioactivity. Lastly, the shifting waters of the tides are raised primarily by the Moon; the Sun contributes only about one third to the height of the tides. All these energies combined — great though they are — comprise only a small part of the vast display of energy before us. The rays of the Sun rule.

The amount of radiant energy that falls upon Earth every day is stupendous, yet it is only an infinitesimal fraction of the entire energy which the Sun pours out into space each second. Seen from the Sun, the Earth is only a tiny point — the size of a dime seen from a distance of 700 feet. Thus, of all the vast energy sent forth by the Sun, in all directions, very little hits our planet; practically all of it is lost somewhere out in the awesome depths of space.

The Sun keeps a gigantic volume of space filled with its blinding brilliance and searing heat. Throughout a much greater volume, the Sun is seen as by far the most brilliant star in the sky. The Sun remains visible as a weaker star throughout a volume of space that extends to distances beyond comprehension. Yet so far as man is concerned, practically all this tremendous amount of energy is wasted, and it has been wasted since the birth of the Sun.

How small a fraction of the Sun's energy is the share of Earth?

Assume that the Sun supplies us with radiant energy not in a steady stream but in single bursts lasting only one second each. The Sun could then deliver Earth's quota of energy by turning the heat on for just a single second *once* every *seventy years.* Of course, we are glad not to get the Sun's heat in such a burst.

It would instantly evaporate Earth, despite the 93 million miles between us.

The reason why such a tremendous output of energy is permanently given off by the sun is that its surface is very hot. When any body is brought to a high temperature, it radiates away considerable amounts of energy, and the output increases steeply with rising temperatures. Even the human body gives off energy by radiation: we lose energy at the rate of 3 ten-thousandths of one horsepower from each square inch of skin surface. The energy output of a glowing heap of charcoal is much greater, because its temperature is so much higher: one tenth of one horsepower from every square inch. This is a considerable flow of energy, so that many a picnic cook has roasted his fingers instead of the frankfurters.

How about the Sun? Its temperature is over 10,000 degrees F. at the surface, and it is capable of supplying energy at the rate of 54 horsepower per square inch. During 1960 the total production of electric power in the United States was about 800 billion kilowatt-hours. This yearly energy output could be equaled by a section of the Sun's surface about as big as ten tennis courts.

The reason for this terrific continuous release of radiant energy from the Sun is the sustained extreme temperature of its surface. If we pull a piece of white-hot iron from the fire, it takes only a fraction of a minute for the piece to cool off considerably. Its color soon becomes cherry, and after a short while it is only a dull red. After a minute or two it no longer glows at all. Why, then, does the Sun not cool off, although it radiates away heat at such a tremendous rate? The explanation, of course, is that the energy losses at the surface are continuously replenished by a steady flow from within.

Now, the fire that could keep the solar powerhouse running as it does couldn't possibly be of the nature of a coal fire. This was clear even a century ago, when the tremendous energy output of the Sun was first put down in solid scientific figures. It was calculated that the Sun as an ordinary "fire" couldn't last long even if it were kept going by burning high-grade coal in pure oxygen – one of the best combustible mixtures known. The Sun as an ordinary fire would have burned to a dead heap of ashes even during the time of recorded history. On the contrary, we know that the Sun must be much older than a few thousand years. It has been pouring out energy at the present terrific rate for *billions* of years – more than a million times longer than a star made of pure coal or gasoline could burn. In the interior of the Sun there must be, buried under thick layers of hot gases, a stupendous reservoir of energy.

To learn something about the interior of the Sun's mighty bulk appears hopeless. How could we tell what is hidden beneath shells of hot gases hundreds of thousands of miles thick – we who are removed almost a hundred million miles from even the outward scene of action? It is a great tribute to physicists and astronomers that, against such odds, they were able to "calculate their way" into the Sun's deep core.

What led to our knowledge of the core is the simple fact that the Sun is obviously a stable structure. It neither explodes nor collapses under its own tremendous weight. Weighing as much as 330,000 Earths, this monstrous mass of the Sun is mostly hydrogen and helium gas, held by gravity in the form of a huge sphere. But what keeps the Sun from collapsing? After all, gases are compressible, and the innermost core must be under a terrific load because almost all of the Sun's mass rests upon it. We should expect the interior to yield to these loads and be compressed solid. This, however, is not the case. A highly compressed body of gas exerts great pressure itself and thus can carry great loads.

An automobile is actually carried by the compressed air in

its tires: it literally moves on air. Furthermore, the carrying capacity of the compressed mass of gases can be increased. We can increase the density of the air inside the tire by pumping more air into the confined space of the innertube; or, we can increase the temperature of the air in the tire. Either procedure will increase the inside pressure — and it is the "pounds per square inch" in the tire that determines the carrying capacity.

With this automobile-tire approach, physicists and astronomers were able to learn something about the Sun's interior. Obviously each shell of gases of which the Sun is built must be able to carry the whole weight of the shells above. This is possible only if the pressure within each shell has the proper value — otherwise it would collapse under the weight above. But the pressure below must not be too large, either, because then the shell would lift the weight above and blow up the Sun. The pressure must be just right *throughout* the Sun. All the astronomers had to do was to assign each shell an appropriate density and proper temperature so that the result is a stable Sun — a Sun that neither collapses nor blows up.

There is only one difficulty in this fascinating game of Sun-building, and the example of the automobile tire shows it clearly. We can get enough carrying capacity from the tires with a low density of air if the temperature is high; but we can also make it at lower temperatures by simply adding air to increase the density. By the same token, astronomers did not know how dense and how hot each shell of the Sun's interior had to be. If they assumed a high density, the temperature could be lower; conversely, if they wanted to get away with a low density for a particular shell, all they had to do was assign to it a higher temperature.

There was, however, only one solution to the problem, as the existence of the Sun clearly demonstrates. The solution was found because we know how much material there is in the Sun.

Progressing from one shell to the next, astronomers had to assume densities in certain steps so that finally, when they reached the center, all of the Sun's material was accounted for. If they started out with too steep a density increase, they used up too much material in the outer layers and finally, when they reached the center, there was no material left to fill it. Or, if they went more slowly in assigning densities to the various shells, they had some leftover material. So there was only one solution, and it had an astounding result for the temperature of the Sun's core: 38 million degrees F.

The most interesting part of this figure is that hydrogen gas starts to "burn" at this temperature and gives off tremendous amounts of radiant energy. But this is a fire of a particular kind. It is nothing of the sort we observe in a stove, a blazing pile of wood, or a gasoline fire. Hydrogen heated to millions of degrees burns in the devastating blaze of an atomic fire.

At temperatures this high, the nuclei of hydrogen atoms begin to fuse, or stick together, and heavier nuclei are formed from the simpler particles that make up the hydrogen nuclei. Every time two nuclei fuse, they give off a flash of radiation that carries a great amount of energy. The principal "ash" of this fearful atomic fire is helium. Deep in the core of the Sun, hydrogen is transformed into helium, and a terrific amount of radiant energy is released. It takes temperatures of many millions of degrees to release energy from matter.

More than fifty years ago the famous physicist Albert Einstein showed that matter is actually highly concentrated energy. In an ounce of matter are hidden tremendous amounts of energy. If all of this energy could be liberated, it would be equivalent to the heat produced by burning more than half a million tons of coal. For this it would be necessary to transform the entire ounce of matter into radiant energy. When hydrogen is transformed into helium in the terrific heat of the Sun's core, some of the matter involved is indeed lost. When 1,000 pounds of hydrogen gas are transformed into helium, the weight of helium produced is only 992 pounds – 8 pounds of matter having been transformed into radiation.

From the hot atomic broiler in the Sun's center, the stream of energy forces its way through the mighty bulk of the Sun in all directions. Finally the flow of radiation reaches the surface, where it replenishes the energy which the topmost layer of the Sun is losing continuously into space.

Six hundred million tons of hydrogen are used up every second; but almost all this reappears in the form of helium. Almost five million tons of matter are truly lost every second. These five million tons are transformed into radiation and spilled out into space. But the mass of the Sun is so great that only one and a half percent of it has been used up since the Earth was born. Regardless of the waste of energy at a rate beyond human comprehension, the Sun has safely stored away a supply to last for tens of billions of years to come. Man need not fear that the Sun will stop shining while he is around.

In the hot core, the very same kind of fire is burning as that from which the hydrogen bomb draws its dreadful power. When an ordinary uranium or plutonium bomb explodes, enormous temperatures are created in the very center of the blast. These

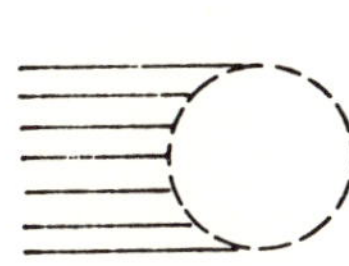

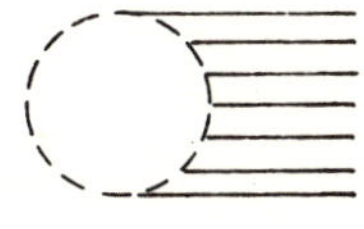

temperatures are so high that they can cause a mass of hydrogen to fuse into helium and to give off its terrific power in the process. In this way the awesome H-bomb explosion is produced, surpassing the violence of an ordinary atomic bomb by a thousand. The old-style bomb serves only as a trigger — once the hydrogen is hot enough, it explodes by itself. Because hydrogen nuclei are transformed into helium and energy only under conditions of extreme temperatures, the atomic physicist calls a hydrogen bomb a "thermonuclear weapon." Fortunately, in an H-bomb explosion the temperature drops within a very small fraction of a second, and the terrific explosion of hydrogen comes to a dead stop at the instant it loses its terrific heat. There is nothing to keep the heat together; the violent explosion pushes everything aside.

The core of the Sun is a constantly burning atomic fire equivalent to the center of an exploding H-bomb. In the core the hydrogen keeps on burning at these tremendous temperatures, since the heat cannot escape in one violent explosion — it can only trickle out slowly. The enormous weight of the Sun's mass upon the core keeps the atomic furnace together all the time. If a fistful of this fearfully hot gas from the Sun's core were suddenly lifted out from its prison, it would immediately explode with the devastating violence of an H-bomb.

Confined by heavy, thick gas walls, the savage core of the Sun turns out atomic energy in gigantic quantities. The energy forces its way through the deep outer layers and leaks out into space. Thinning out through the great distance between the Sun and the Earth, the rays of the Sun finally fall on our planet. There is nothing in the friendly light to remind us of the awesome, devastating heat of the place that engulfs the birthplace of the Sun's energy.

6 • *The Music of the Spheres*

One of the grandest dramas known to man is the march of the planets of our solar system along their orbits. The drama is in constant performance before our eyes, with the Sun as the leading character, and the planets, including our own Earth and its Moon, in supporting roles. Man has long observed this drama almost as he might observe a play. Unlike man-made plays, this one has no certain beginning or end, and even lacks any evident plot. But these stirring events challenge human curiosity, and the history of our efforts to understand them reaches back deep into the past.

For thousands of years, celestial objects and their behavior were understood mainly in poetic and religious terms. The Sun and the Moon were gods, and the heavens were their stamping ground. When, a mere three hundred years ago, Kepler and Newton grasped the laws of planetary motion, man became a spectator with a basic understanding of the spectacle. Then, in our own century, the spectator himself began to take an active part, building his artificial moons and interplanetary missiles.

Many of us have seen an artificial satellite glide silently across the evening sky and felt deeply moved, even shocked, by man's successful intrusion into this realm that was for so long forbidden to him. At the same time we have wondered about the strange path that a satellite weaves across the skies of the whole world. Artificial satellites, once they enter upon the heavenly stage, follow the rules of planetary and lunar motion with utmost exactitude. Man, long a mere spectator of nature's show, had to master the rules of the script before he could put on a show of his own.

The actions of the heaven's players have been much the same since the creation of the solar system some four billion years ago. During nearly all this awesome length of time the action went on with no intelligent observer around to watch it. Not

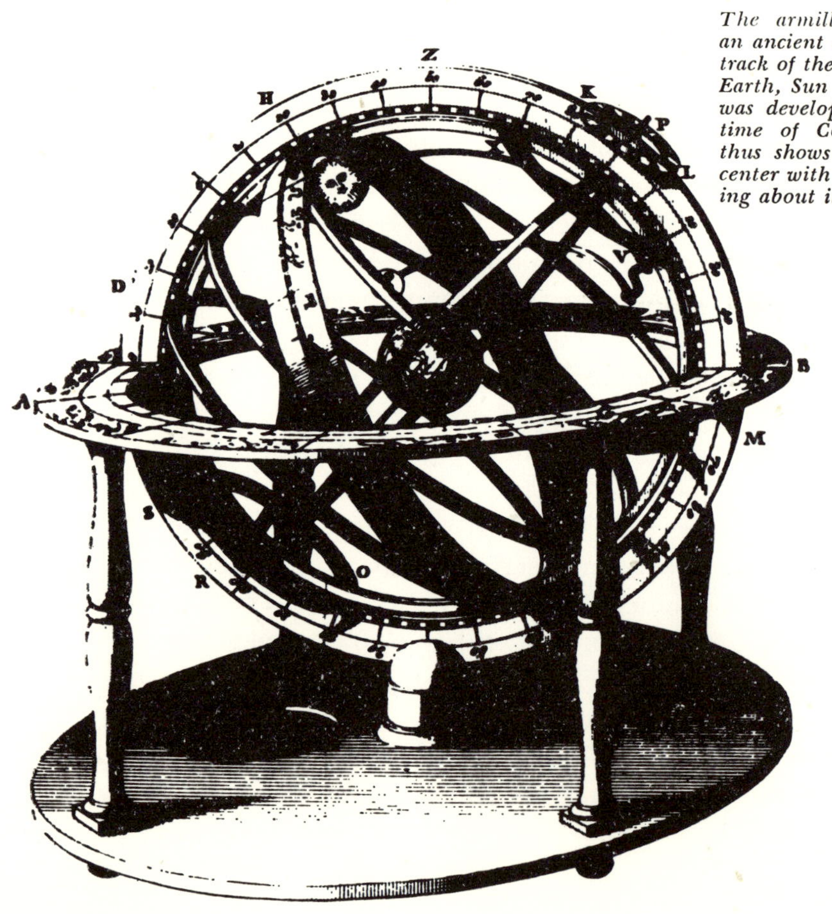

The armillary sphere is an ancient device to keep track of the movements of Earth, Sun and Moon. It was developed before the time of Copernicus and thus shows Earth in the center with the Sun rotating about it.

until 10,000 to 15,000 years ago, perhaps, did man become sophisticated enough to attempt interpretations of the great play – interpretations reflected later in the ancient cultures of the Chinese, the Indians, the Mayas, the Egyptians, the Babylonians. These ancient peoples not only tried to understand the motions of the stars; they attempted the much bigger task of giving an account of how the whole universe came into being. They solved this task with vivid religious imagination, and the most picturesque myths have been handed down through the ages of how the gods created the world.

In the sixth century B.C. the great Greek philosopher Pythagoras and his disciples broke the spell of myth and made the first rational attempt to understand the spectacle of the sky. They had the brilliant notion that the motions of the lights in the sky could be understood only through mathematics – through numbers and geometry. Looking upon numbers as the essence of all things, they had only contempt for the vulgar merchants who used numbers merely to count their wares and to compute their profits. Pythagoras taught that numbers are the key to the understanding of nature and the universe. This idea is still very much alive today; in fact, it was the beginning of our modern scientific thinking. When the Pythagoreans applied their mathematics to the sky, they formulated the first scientific interpretation of the heavenly drama.

To them, the Sun, the Moon, and the planets appeared to revolve around Earth, which appeared to be at the center of all things. They explained these movements by the existence of eight concentric spheres, one each for the Sun, the Moon, and the five visible planets (Mercury, Venus, Mars, Jupiter, and Saturn), the eighth sphere being the one that bore the fixed stars. Earth was at the center, and the spheres, with the various heavenly luminaries fastened to them, slowly turned round and round, each with its ordained speed.

A certain size was assigned to each sphere, and here the Pythagoreans applied their keen sense of numerical harmony. One of their most cherished possessions was, indeed, discovery of the mathematical law of harmony made by their master. Pythagoras had found that musical chords are produced by strings whose lengths represent a series of simple numerical intervals; and so the Pythagoreans made the mathematical laws of music an integral part of their world system. They assumed that the size of the planetary spheres increases from one planet to the next in the same fashion as the length of strings increases in a musical instrument such as a harp. The heavenly spheres, however, were thought to be made of the clearest crystal, and as they swung around in their majestic motion the glassy bells rubbed against one another and brought forth a harmonious music, so heavenly and sweet that it was perceptible only to the ears of the gods. This was the famous harmony of the spheres.

The Pythagorean interpretation of the heavenly action is too poetic for modern scientific thinking. Yet it was much more matter-of-fact than a view of the planets as gods that go about their fickle and inscrutable ways. For the first time the motion of the stars was described in mathematical terms.

Only two and a half centuries later, the Pythagorean way led to a correct solution of the heavenly riddle — that of the Greek astronomer Aristarchus of Samos, in 280 B.C. He claimed that the Sun is in the center, while the planets, including our own Earth, wheel around the central fire along huge orbits. Aristarchus was thus the first to picture clearly and correctly the grand design. It was a tremendous intellectual achievement, but the world was not yet mature enough to grasp and to hold it. The great thinker's vision was forgotten for almost 2,000 years — all through late antiquity and through the Middle Ages. In the mind of man Earth still stood in the center of the universe; nature's magnificent spectacle remained misunderstood.

The medieval notions of the heavens were finally swept aside by the clear thinkers of the Renaissance. The critical year was 1543, when Nicolaus Copernicus published his momentous work *On the Revolutions of the Heavenly Spheres.* Publication took place after his death because he did not dare to voice during his lifetime views that were so contrary to the "absolute truth" as seen by the church fathers and laymen alike. Like Aristarchus almost twenty centuries before him, Copernicus pictured the Sun in the center of the family of planets. This time, however, the idea took hold and marked the beginning of a great revolution in the thinking of man. The broad pattern of the heavenly play was finally glimpsed. But this was only the beginning of understanding; the finer points were still obscure.

To the astronomers of old, it went without saying that the planets move in circles. Only a mathematically perfect circle with its marvelous harmony was considered fit to serve as an orbit for a heavenly body. Even Copernicus was convinced that the orbits of the planets and moons are true mathematical circles. But, as it turned out, the motions of the planets did not quite fit into this man-made scheme of circles; nature's mathematics was more complex than man supposed. It was Johannes Kepler of Germany, born three decades after Copernicus' death, who first satisfactorily explained planetary motion. His understanding was gleaned from long, detailed, exact observations of the heavens made by Tycho Brahe and others.

In the first of his famous laws, Kepler stated that all celestial orbits are ellipses, not circles. The shape of an ellipse is hard to describe; it has to be seen to be understood and appreciated. An ellipse looks somewhat like a circle that has been squashed from two opposite sides, but it is more graceful than this description implies. In a circle the curvature is the same all around, while the curvature of an ellipse changes gradually from point to point. An ellipse has a center, but its diameter differs in different directions. The longest diameter is called the major axis; at right angles to the major axis is the shortest diameter, called the minor axis.

Each ellipse has two special points located on the major axis at equal distances from the center. These two points – the foci, or focal points – display a strange geometrical law of the ellipse. Choose any point on the circumference of the ellipse and from it draw a line to each focal point; now do the same from any other point on the circumference. Measurement will show that the total length of the two lines from each point on the circumference is the same as the total length of the lines from the other point. The same holds for any two points on the circumference of an ellipse; in fact, it is in just this way that an ellipse is mathematically defined.

After this brief excursion into the geometry of ellipses, we return to Kepler and his first law. It states in full: "The orbit of each planet is an ellipse, with the Sun at one of the focal points." The same law holds for the orbits of moons, which swing around their planets in elliptic orbits: in each case the

The elliptical path of a planet around the Sun.

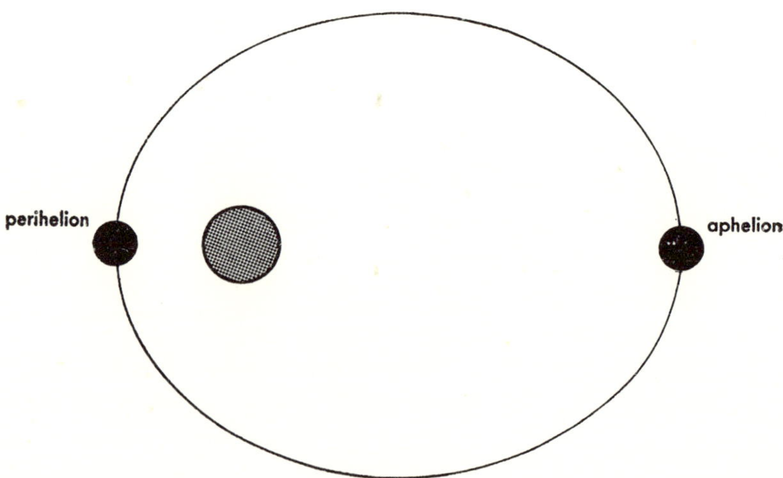

center of the respective planet coincides with one focal point of the lunar orbit.

The orbit of Earth around the Sun is an ellipse that differs only slightly from a true circle. (The same is true for the orbits of the other planets, and this is why they were so long thought to be circles.) Being elliptical, the orbit of Earth is somewhat off center with respect to the Sun, and Earth gradually changes its distance from the Sun as it swings around. The point where Earth is closest to the Sun is called the perihelion (Greek *peri,* "near," and *helios,* "Sun"). The point where Earth is furthest from the Sun is called the aphelion (Greek *apo,* "away from").

Kepler's first law explains the shapes of celestial orbits; his second law has to do with the speed of celestial bodies on their repeated round trips. It proved the old astronomers wrong who believed that the planets and moons roll along serenely at a constant rate of speed. Kepler found that planets and moons change their speed constantly. Earth, for example, is moving fastest as it passes through the perihelion of its orbit; then it gradually loses speed until, at aphelion, it is slowest; and then it picks up speed again as it heads toward perihelion. Planets move around the Sun somewhat like a yo-yo leisurely swinging around in a vertical circle, slowly when it sails overhead, and rapidly when it zips through the bottom part of the "orbit." Kepler's second law describes this ever-changing speed in mathematical terms.

Kepler's laws were a triumph for man: the spectator now understood at least part of nature's play on the stage of the heavens: that is, *how* the planets move. But there was still one big question: *Why* do the planets move as they do? The man who found this answer was born in 1643, exactly one hundred years after the death of Copernicus and the publication of his famous work. The man was Sir Isaac Newton of England, and his discovery was nothing less than the master principle of the universe: the law of gravity.

The story is that Newton, sitting under an apple tree, saw an apple fall. That gave him an inspired thought: the very same force that makes apples fall also keeps the planets and moons in their orbits. This anecdote is probably an invention, completely; but Voltaire, the witty Frenchman of the eighteenth century, remarked that this story would have had to have been invented if it weren't true. As a great admirer of Newton, Voltaire perhaps invented the story himself. But no matter; Newton conceived the idea of gravitation and formulated its law, and so became one of the giants of human history.

This is not the place to discuss Newton's law in all its mathematical brilliance; we shall only follow his reasoning to the

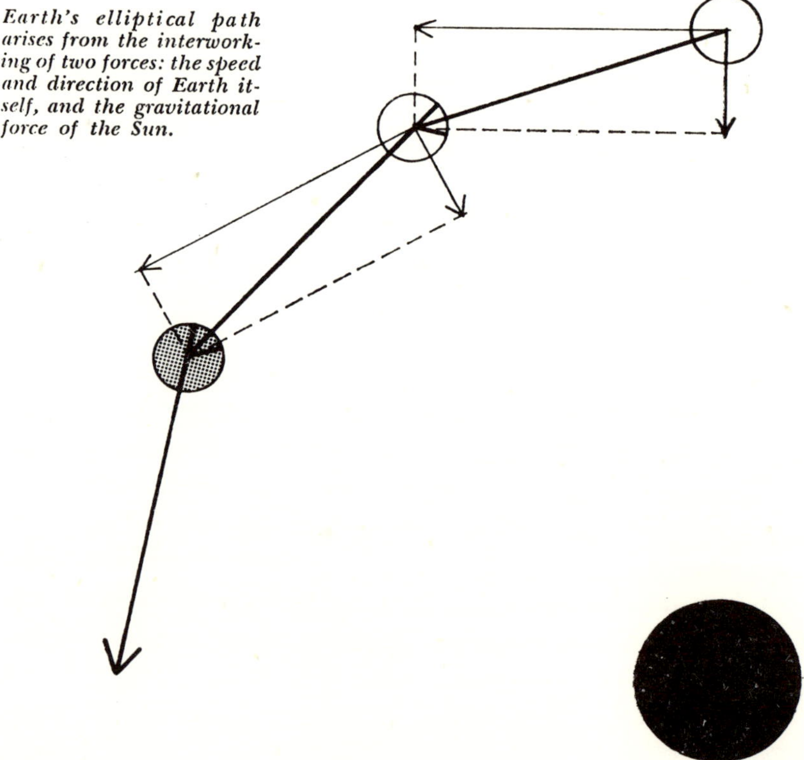

Earth's elliptical path arises from the interworking of two forces: the speed and direction of Earth itself, and the gravitational force of the Sun.

point where we can see why the planets and moons move as they do. Also, we want to make it clear to ourselves that without the concept first uttered by Newton we should not know how to fling artificial satellites into their globe-spanning orbits.

Gravity is the force of attraction that exists between all bodies in the universe: suns, planets, moons — and apples. Newton showed that this force can be calculated if we know the masses (amounts of matter) in bodies and the distances between them.

Let us see how gravity affects Earth and the Sun. A tremendous force of attraction acts between these two bodies, but Earth, because of its motion, is able to resist this constant tug. If Earth suddenly stopped moving in its orbit, it would start falling toward (that is, being drawn toward) the Sun, and after about two months of increasingly rapid fall it would be engulfed. Conversely, if the Sun were suddenly removed from the scene, Earth would immediately cease moving in its elliptical orbit: it would move along a straight line out into the depths of space. As things are, Earth's orbit is the result of the inertia of the planet's motion and the force of gravity working simultaneously. Earth is constantly pulled sideways, so to speak, and its flight path is thus a closed orbit.

Newton, then, used gravity to derive Kepler's laws of planetary motion. His formula took account of the law of gravity, the masses of the Sun and planets, the distances between the bodies, and the times of revolution. It was now possible to show mathematically how billions and billions of tons of matter, balled together in the planets and their satellites, moves through millions and millions of miles of space along precise paths.

An admirer of Newton's genius once said: "Newton is indeed a most fortunate man; there is but one universe, and so it was given to but one man to discover its master principle."

Today when rocket engineers fling their artificial satellites into the sky, they follow the rules that Kepler and Newton discovered. Small though artificial satellites may be, their movements around Earth are like Earth's movements around the Sun. The orbit of every artificial satellite is an ellipse.

To launch a satellite into an orbit several hundred miles out, a speed of more than 17,000 miles per hour is required, because the gravity of Earth is very strong. A satellite needs a great push to balance this strong force; it takes rockets with three or four stages to create this push. The trick is to guide each rocket stage in such fashion that the satellite is injected into its orbit at the right height, the right speed, and in the right direction. As Newton's equations tell us, these are the three conditions that determine exactly what orbit the satellite is going to follow.

Satellite launching rockets have to be controlled with great accuracy, because the shape, size, and position of the resulting satellite orbits are very sensitive to even small changes in height, speed, and aim at the point of injection. Some satellite launchings are unsuccessful even if all rocket stages perform to the end. Often the satellite is not aimed in the proper direction

Diagram of a sub-orbital (left) and orbital (right) space flight.

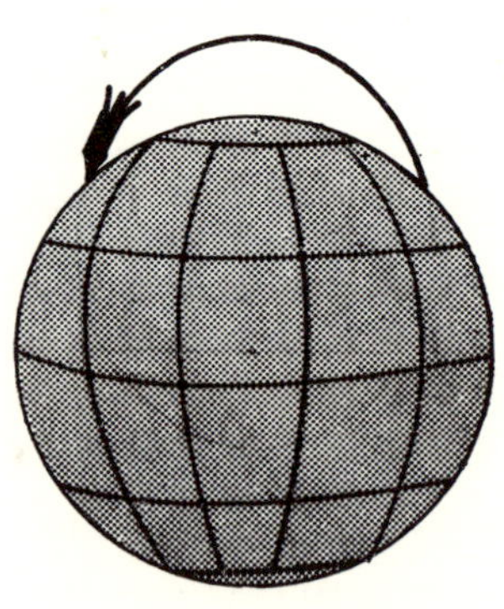

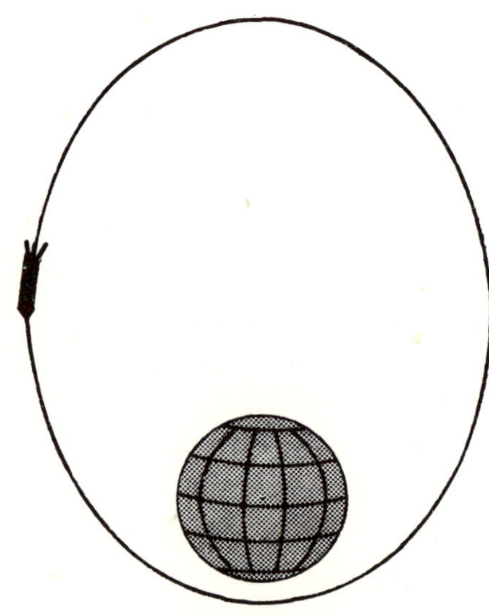

or has not been given enough speed; hence an orbit results that leads the satellite back into the denser layers of the atmosphere. Then it is destroyed like a flaming meteor in the terrific heat of friction.

One focal point of the elliptical orbit of a satellite lies at Earth's center, so that the position of the orbit is always somewhat off-center relative to the globe. The point where the orbit is closest to Earth is called the perigee (Greek *peri,* "near," and *ge,* "Earth"). Exactly opposite lies the farthest point, the apogee (Greek *apo,* "away from").

Consider a special example of the motion of a satellite that would come about under specific conditions: A satellite is injected into orbit at a height of 300 miles above Earth in a direction at right angles to the vertical. At 300 miles the resistance of the atmosphere is practically negligible. The speed of the satellite at the point of injection may be exactly 19,000 miles per hour; this speed is greater than that required to offset gravity. Serenely the satellite rises along its arc, steadily losing speed because of Earth's gravity. Finally it reaches the highest point of its orbit, the apogee, at a height of about 1,400 miles, exactly opposite the point of injection. Then the satellite sweeps down along the descending arc of its elliptical orbit and, according to the second law of Kepler, it picks up speed. It rounds Earth, continuing to gain speed, and as it reaches the point of injection 300 miles above the ground (this is the perigee) it is going at 19,000 miles per hour as before. Then the satellite begins its second round.

The first satellites launched by Russia and the United States entered their orbits at levels lower than 300 miles. Here there was enough atmosphere to cause a small but appreciable amount of friction. As the satellites moved around Earth, trip after trip, this small friction made itself felt and ate away the momentum of the satellites, so that their orbits shrank slowly and the satellites spiraled back toward Earth. After a lifetime of a few weeks or months a satellite would enter increasingly dense layers of the atmosphere and burn up.

The first successful Vanguard satellite, about the size of a grapefruit, follows a particularly wide orbit. Its perigee lies over 300 miles above the ground, beyond the pernicious drag of the atmosphere. This little satellite is expected to swing around the earth for more than a century. It will outlive all people that were alive when the tiny ball was elevated to the rank of a celestial body taking part in nature's drama.

In principle, there is little about the motion of heavenly bodies that the founders of celestial mechanics, Kepler and Newton, did not know. In their time man had not yet put electricity to use, and the steam engine was still more than a century in the future. Kepler and Newton used goose quills to write down the laws of nature, and did so by the light of candles. Yet these mental titans laid the foundations of astronomy, and with them began the most exciting technical adventures of modern times: the conquest of space.

7 • *A Moon Is Born*

A meteoroid floats through space. It is about the size of a fist — an irregular chunk of metal with a rough surface, weighing perhaps 20 pounds. The blinding rays of the Sun hit it like a spotlight, and it glitters against black space like a dust mote in a darkened theater. Slowly it tumbles about its own axis, and an ever-changing pattern of sharp lights and deep shadows plays over its surface.

Were it not for the slow rotation, one would think the meteoroid was suspended in space, motionless. Actually, it hurtles forward with a speed of many miles in every second, sweeping along a tremendous orbit that leads all the way around the Sun. Since the beginning of our solar system this meteoroid has swung around the Sun billions of times, like the planets in their own serene orbits. Its vast trajectory crosses the orbits of Venus, Earth, and Mars. During the long journey of the meteoroid there may have been thousands of near misses between it and these three planets.

Now the meteoroid is approaching Earth. If an observer were moving beside it, he would see the disk of the planet slowly growing larger, brighter against the blackness of space. The meteoroid is on a collision course, approaching the planet from the rear with its superior speed.

The pull of Earth's gravity now adds to the speed of the meteoroid. Earth's disk grows still larger and larger — now it spans half the sky. A few atoms of air begin to impinge on the leading face of the meteoroid and tear away the fine cosmic dust accumulated there during the eons of the past. Seconds later the object — now a visible meteor — begins to glow as it rams down into the denser air below. The terrific friction produces a heat of thousands of degrees, and the meteor becomes a streak of luminous metallic vapors and incandescent air. Minutes later a softly glowing band still marks the path where the meteor met its end. Earth, with its atmosphere, has gained about 20 pounds in mass.

Every hour Earth's air engulfs uncounted millions of meteoroids, most of them microscopically small. Their number is so great that the planet grows by about 1,000 tons every day. Many scientists believe Earth and the other planets originally formed from billions of small particles—meteoritic material—that filled the space around the Sun and gradually accumulated to form larger masses. The formation of Earth is not yet fully completed, and the fall of our meteor was just one more event among billions that once brought Earth into existence.

Whatever comes in from outer space Earth holds fast with the mighty grip of her gravity. All meteoritic material that has ever come within this grip is still part of Earth today. Except for a few forlorn atoms that escape into space from the topmost layer of air, the planet has never given up a single piece of matter once gained. Gravity is a jealous, efficient guard.

Just as our meteor plunges to its flaming death, a celestial body of another kind enters the scene. As it emerges from the shadow of our planet, it reflects the sunlight sharply. Like the meteoroid, it hurtles through space and it consists of metal; but in other respects it is strikingly different. It is regular in shape — a perfect sphere with a highly polished metallic surface that mirrors the Sun in a blinding highlight. It also weighs about 20 pounds. But this 20 pounds makes up one of the most remarkable pieces of matter in the physical history of the planet: the artificial satellites. This one was, like the others, once part of Earth's mass, but it has just left Earth to begin its existence as a celestial body equal in rank to the Moon.

From the beginning of Earth's history, the traffic of matter near it has been one-way. Now, for the very first time, pieces of solid matter are traveling in the opposite direction – *away* from Earth.

This reversal of traffic in the vicinity of Earth could only happen through man's technical ingenuity. He found a way of overcoming the force of gravity that normally keeps everything chained to Earth. The thrust of rocket engines sends satellites away from the planet and lifts them into an orbit around it. The satellites are moons of man's own making.

Unlike the natural Moon that lovers have gazed at through the ages, the artificial satellites are intricate, highly sophisticated tools of science. Even as an inert shell of metal rounding Earth, a satellite becomes a revolutionary tool that can solve many problems for which there was no answer before. But scientists are more ambitious: they are making satellites that function as alert, accurate observers. In many ways satellites act as though alive. They are equipped to "see," "hear," and "feel" what is going on in the fringe area between atmosphere and space. They have an electronic "memory" to remember their "sensations," and a radio voice capable of giving answers to questions radioed from the ground.

For a satellite there is a lot to see, hear, and feel that can never be observed by an Earth-bound instrument. Some of its artificial eyes keep watch on the Sun and scan the solar ultraviolet rays. Others are trained on Earth below, observing the ever-changing cloud cover. Artificial ears listen to the fine clicks caused by the impacts of tiny meteoroids upon the satellite's metal surface, and other "sense organs" keep a record of larger meteoroids big enough to punch a hole through the hull. Heat-sensitive elements measure temperatures. A "magnetic" sense traces the magnetic field of Earth, and there are devices to detect charged particles given off by the Sun and to record cosmic rays.

Whatever the satellite learns is stored in its electronic memory — a tiny set of magnetic cores or, in some models, a miniature tape recorder taken aloft. Once during each round trip, when the satellite is at the right spot, a radio signal from the ground commands it to spill its stored information. The satellite then speaks through its radio voice, transmitting to the ground all the data picked up during the last revolution.

Satellite-borne instrumentation is a marvel of scientific and engineering design. Each instrument is almost unbelievably light in weight, because an ounce saved in construction may spell the difference between success and failure in getting the satellite into the desired orbit. If a special instrument cannot be built within a strict weight limit, it cannot go along for the ride. Scientists have worked out a highly sophisticated program for these "inboard" experiments that will give the most information for the least weight. Army, Navy, Air Force, several universities, and industries are involved in these experiments.

In the first satellites the batteries for the power supply claimed the lion's share of weight — about one third of the entire weight of the satellite. The rest of the weight went into the electronic gear: receivers, transmitters, relays, multi-channel telemeter, and memory units. These items are so tiny that they put the two-way wrist radios of cosmic-strip supermen to shame. All the miniature marvels of the modern electronic art go into these sets: transistors, printed circuits, tiny condensers, magnetic cores — so tiny and light that everything can be crammed into a package no larger than a stack of half a dozen saucers. Each of these electronic units has more individual parts than a big color television set, yet it fits into the palm of one's hand. The electronic equipment of a satellite can perform the most intricate tasks of sensing, measuring, storing and transmitting at a minimum of power expense. Small batteries can run the equipment for one to two weeks.

Some of the more recent satellites are equipped with solar batteries. These promising instruments of the modern electronic age are able to convert an appreciable percentage of solar energy into useful electrical power. Solar batteries are most ideal for satellites, since there is an abundance of free sunshine in space. Unless a meteoroid hits a satellite and destroys the delicate circuitry, space-borne radio and TV equipment powered by solar batteries can virtually run forever.

Since the first Sputnik started on its lonely trip into space, the population of satellites has been steadily increasing even though some of them orbited only for a few weeks or months. At one time – in January of 1962 – a total of 35 man-made vehicles were coasting through space, many of them still busy relaying their scientific measurements back to their creators on Earth. Not all of these vehicles are true satellites. Russian engineers have made two successful attacks on the Moon. On September 14, 1959, at 5 p.m. New York time, the first object originating from Earth reached the Moon. A spherical instrument package with a total weight of 858 pounds crashed on the lunar surface with a speed of 7,500 miles per hour after a trip of about 35 hours. On October 6, 1959, a Soviet rocket, launched three days earlier, coasted around the Moon and upon radio command from Earth pointed its two cameras at the back of the Moon. An intricate mechanism developed the pictures inside the rocket, scanned them, and radioed them to Earth, affording a glimpse of lunar landscapes that have so far been hidden from the searching eye of man. The photographs revealed about 70 per cent of the invisible part of the Moon's surface. The Soviet scientists exercised the classic right of a discoverer by naming several conspicuous craters, mountain ranges, and plains.

Both American and Russian space scientists have launched rockets into interplanetary space by imparting to them a speed great enough for them to break free from Earth's gravity and enter an orbit around the Sun. Of these so-called planetary probes the U. S. "Pioneer V" has so far been the most rewarding one. The signals from its 150-watt transmitter were received up to a distance of 13.3 million miles from Earth. Although this transmitter was expected to maintain contact over a distance of 50 million miles, it failed, probably because of a storage-battery leak.

Since the beginning of the satellite age, the weight of the space-borne vehicles has grown from a few pounds to the 2½-ton "Midas" satellite of the U. S. Air Force, and the Russian satellites weighing over five tons. These vehicles truly deserve the name of "space ship." However, the most significant difference between a cannon shell, a rocket, and a ship is that the latter returns whence it came – more or less intact. In space flight the problem of re-entry into Earth's atmosphere and of final recovery is beset by fantastic difficulties. We have seen that a body is able to become a satellite only by virtue of its great speed, which must exceed the critical limit of about 17,000 miles per hour. Before a satellite can be recovered this enormous speed must be killed. As yet no space vehicle can carry enough fuel to exactly reverse the process of launching, namely, using rocket power to brake all of the speed. Most of the satellite's momentum must be traded off against heat of friction as the vehicle lashes through the atmosphere. The chief problem, of course, is to prevent a returning space craft from ending as a flaming meteor.

American space scientists instituted the "Discoverer" satellite series, in which they attempted to retrieve a 300-pound instrument capsule ejected from a satellite previously put in orbit. The first 12 trials failed, until on August 11, 1960, the capsule of the Discoverer XIII satellite was fished from the Pacific by a Navy helicopter pilot. The capsule had circled Earth 17 times and was shot back into the atmosphere by radio command from the ground. This was the first time that an object had been recovered after a genuine trip into space. A few weeks later the Russians succeeded in retrieving two small dogs — the first higher forms of life ever to have survived a trip into space. These canine space travelers were followed in the spring and summer of 1961 by the first human astronauts in history, Majors Gagarin and Titov of Russia.

The American counterpart of the Russian experiments in manned space flight is Project Mercury of the National Aeronautical and Space Administration. Since the classic flights of Commander Alan Shepard, Major Virgil Grissom and Lieutenant Colonel John Glenn, the world public has been thoroughly familiarized with the shape of the Mercury space craft.

The astronaut occupies a round, tapered metal capsule which has a flat conical top. Although the operation of the rocket boosters and of the satellite capsule itself is fully automatized, the astronauts are able to control the capsule's attitude and to operate all flight controls such as the firing of the retro-rockets that initiate the most critical phase of man's trip into space — the safe descent back to Earth. The capsule is designed to avoid most of the heating by atmospheric friction, while a parachute system lowers the capsule to a safe landing into the ocean.

Paralleling the technical development of the intricate Mercury machinery, the training of seven astronauts has been in operation since the summer of 1959. The actual launching of these American manned satellites from the sand dunes of Cape

Canaveral in Florida is a most dramatic event; it is the climax of thousands of hours of work of many trained minds and skilled hands. The actions of man and machines must blend in a perfectly controlled, well-timed operation. This operation is like a huge jigsaw puzzle with the pieces fashioned to fit accurately; each must fall into place at a precise instant. It is the kind of operation in which American engineering managements excels.

The time is early morning. During the night, the huge missile test center at Patrick Air Force Base has buzzed with activity. The satellite launcher is still surrounded by the gantry, a steel framework that permits access to all stations of the rocket combine, which is many stories high, one above another. Engineers and technicians are checking and rechecking the many functional parts of the complex machine – its delicate "brain," its electrical "nervous system," its mechanical "muscles." The tons of propellants are safely stored inside the huge tanks, their pent-up chemical energy waiting to be unleashed in fiery union. Thick cables run from the rocket over to the gantry frame, connecting the hundreds of check points, relays, and activators within the big machine with the command post inside a solid concrete blockhouse. The walls of the command post are lined with banks of checklights, red, green, and white; there are automatic plotting boards with wavering pens scribbling their messages, and green line patterns jitter silently across the screens of oscilloscopes.

There are thousands of items on the pre-flight check lists. After the last check mark is made, the command is given to remove the gantry. It rolls aside slowly, leaving the tall rocket perched on its base like a huge obelisk. But the rocket is still on a leash, consisting of thick cables that connect the "brain" of the rocket with the gantry on the side. The Sun is up now, and the mighty machine glitters in its light.

A loudspeaker blares its terse message across the area:

"X-minus two minutes! Minus two minutes."

The tracking stations on the Bahamas and on the string of islands to the southeast have been on duty for hours. They are in direct contact with the launching site via radio and undersea cable.

"Thirty seconds! Rocket pressurized!"

The rocket is now ready for blast-off, and the very last phase of the countdown begins. The monotonous voice of the speakers all over the area wears the nerves. During these few seconds the men who have created this marvelous machine tensely relive the thousands of hours of work, the worries and the sleepless nights. Within seconds all their concentrated efforts will culminate. They think of the brave man tucked away in the tip of the huge rocket. Success or failure – it is a sharply cut proposition now, with no room in the middle.

". . . Two . . . One . . . Zero . . . IGNITION!!!"

A lashing flame breaks out at the bottom of the rocket. It writhes and whips with a blinding glare, then steadies. A bellowing roar engulfs and shakes the area. The stabbing jet of exhaust gases becomes furious, and through the noise one hears a rising whine as the propellant feed pumps of the first stage are whipped up to full speed.

The huge missile trembles. Just as the noise level becomes unbearable, the tall structure begins to rise, slowly, like an elevator. Thin jets from its controls spurt around its base. The rocket is suspended in midair, perched on a column of white-hot gases. Upward it moves, still slowly, the thrust of the exhaust straining against the tremendous weight of the structure. Then it gains speed as the exhaust pushes relentlessly. Faster and faster the rocket rises, trailing a jagged column of vapor. The glare of the exhaust flame fades as the rocket moves away. Now the missile is racing upward like a fiery lance hurled into the sky.

In the blockhouse, busy pens trail across long paper strips. Radar and optical tracking devices follow the rocket's path, plot its trajectory. Data are fed into computers, and the rocket's future course is calculated swiftly.

In the rocket, the engine tilts slightly upon command from the missile's "brain." The flight path of the vehicle is slowly tilted toward the east. Sensitive gyros keep it in balance. All during the time since take-iff the astronaut is pressed into his contour seat by his slowly increasing body weight as the rocket accelerates upward at an increasing rate. Relentlessly he is being whipped forward by the fierce power of the rocket which is now racing through the thin air of the upper atmosphere. The vehicle is on course. According to schedule, the escape tower is being jettisoned and the actual space craft with the man inside is pushed free. The launching operation has reached its climax – the man is in orbit. The burnt-out booster follows lazily, also in an orbit of its own.

During these phases of operation the tracking stations on the ground have been following the course of this new intruder into the realm of space. The island stations are on a line beneath the course: Grand Bahama, San Salvador, Mayaguana, Grand Turk, Puerto Rico, Antigua. Their tracking data are pooled and fed into a central computer, and within a minute or two the computer has the answer, namely the data of the orbit: perigee, apogee, orbit inclination, time of revolution.

Noiselessly the astronaut plunges through the darkness of space. He is now weightless, and were he not strapped down in his seat he would float freely inside his tiny metal box. Until the retro-rockets preparatory to re-entry are fired hours later, the familiar concepts of up and down no longer exist for the astronaut.

Below stretches the vast expanse of the tropical Atlantic. Long strings of puffy clouds gleam in the light of the Sun. They look like little wads of cotton glued to the surface of the Earth. The planet's spherical shape is apparent from the sweeping arc of the horizon. After only twenty minutes of flight the coast of Africa appears below. The continent is set off against the inky sea. The satellite crosses the equator and races toward the Indian Ocean. The sun slips below the horizon behind, and within seconds the astronaut is plunged into darkness — he enters his first "night." The universe beyond is like a curtain of utter black sprayed with steadily shining, needle-sharp stars. The broad band of the Milky Way, now seen through no obscuring atmosphere, is brilliant.

Glittering in the light of the Moon the space craft races over the western coast of Australia, crosses the equator in a northerly course and spans the vast expanse of the Pacific. Barely an hour has passed since blast-off.

About twenty minutes later the space craft approaches the west coast of the American continent. At the same time the sky ahead begins to glow. Minutes later the clouds below begin to light up, and then, in a mere second, the mighty Sun shoots above the horizon. The craft reflects the light like a blinding flame; the solar radiation strikes an almost physical blow.

The continent below is in the grey of early dawn. The black mountains of the coastal ranges are barely visible against the dark grey fog banks. But the space craft already glitters in the light of the Sun, more than 100 miles above the surface of the Earth.

The tiny capsule passes over the California coastline. A little more than three minutes later, it flashes over the 80th meridian, completing its first circuit around the Earth.

The astronaut, of course, is more than just human cargo. All during his trip he is busy with countless technical tasks, performing his mission in space.

The same fantastic trip around the planet is repeated several more times until at an exactly predetermined instant — checked and rechecked many times over — the re-entry operation begins. The retro-rockets kill some of the fearful speed, but enough of it remains to make the descending capsule a man-made meteor that is visible to the crews of the waiting ships below as a huge ball of fire slashing across the sky. When the proper altitude is reached a small drogue chute — an aeronautical sea anchor — is released which is designed to dampen the swinging motion of the capsule. Its speed is finally checked to a mere 25 feet per second by the main parachute, a ring-sail type cargo chute measuring 63 feet across. Upon hitting the water the parachute is immediately detached from the capsule by a small explosive separator charge to prevent drifting of the capsule with the wind. To minimize the time between impact and final recovery, the capsule is equipped with a radio beacon, a dye marker, and a 5-million candlepower flashlight.

The artificial satellites, manned or unmanned, are not mere gilded containers — scientific ornaments or toys into which treasures of scientific ingenuity, human courage, labor, and just plain dollars have been poured. Each satellite with its human observer or its manifold artificial senses in space is a faithful watcher of heavens and Earth, a worker at the very frontiers of science. In the knowledge it can bring back to mankind it is worth its weight in gold many times over.

8 • *Frontiers in Space*

The first artificial satellites have been hailed as man's first stepping stones into space – the first advances across the space frontier. Newsstands, bookshops, and libraries are full of literature depicting the colorful vistas of this frontier. We read of huge manned space stations – whole cities in the sky. Trips to the Moon and the planets, the conquest of the Galaxy, and the colonization of alien planetary systems become commonplace.

"Space operas," of course, do not pretend to be more than entertaining fiction: they are essentially cowboy stories of planetary and galactic dimensions. But there is also the literature of space technology supplied by professional engineers and scientists. In recent years, national governments and industry have become space-minded. They have established research departments where engineers and scientists work on problems of astronautics and space medicine. From their efforts has emerged a fairly clear picture of the potentials of space flight.

Man has already boarded a rocket ship for his first ride through space. Nobody doubts any more that the basic problems of interplanetary flight could be solved one way or another during the next ten to thirty years. We must ask ourselves, accordingly, what business man has in space.

Science fiction has done much to educate, but also to confuse, the public about the true goals of space operations. Even newspaper reports about the space efforts of the United States and Russia often misrepresent the true significance of the news. Its imagination fired by successes to date, the public becomes easy prey to wild promises about exploring and colonizing the planets, or even reaching other stars. It is time to consider.

Take the planets, for instance. Since the time of Galileo and his 30-power telescope, astronomers have amassed detailed data about the physical and chemical make-up of the alien worlds in our solar system. Much is still unknown, but we can say with virtual certainty that none of our neighbor planets has anything to offer man that could be considered as being of great practical benefit. Judging from present knowledge, scientific and economic returns from a large-scale exploration of the planets are unlikely to justify the expenses of public money and scientific talent required. Within the foreseeable future, such projects are to be justified primarily in terms of international prestige.

Important scientific discoveries have been made in the past by resolute exploration of the unknown; hence those who question large-scale exploration are often accused of hampering the progress of science. Still, the cost of adventures into space in terms of money and scientific talent must be honestly measured against the prospective benefits. In the past, the unknown could be explored at rather moderate cost. The equipment used by Faraday to discover electromagnetic induction could be bought easily for a few shillings. Apparently, our forefathers made all the cheap experiments and left the expensive ones to us. Certainly this is true in regard to planetary exploration. What we know about the planets already indicates that exploration of them by direct human contact is unlikely to produce any revolutionary scientific discovery such as those that have changed the course of human history in the past. In terms of human affairs, there are only two really important bodies in the solar system: the Sun, which serves as a stabilizer for the system and as a source of energy; and Earth, with characteristics particularly favorable to life.

It is, after all, a matter of emphasis. Although other planets, perhaps even the stars, are legitimate targets for adventuring mankind, the emphasis on manned space flights is now out of proportion to their real importance. A similar expenditure of money and talent on great human needs such as education, technical aid to underdeveloped countries, and conservation of the planet's resources would bring immediate practical benefits and a kind of international prestige more to be desired than prestige based on stunts.

Yet there *is* a tremendous future in the space frontier. For years to come the prime purposes of space flight can be centered right on Earth itself. Many practical jobs can be assigned to artificial satellites — jobs of great importance in our technical civilization.

For one example, consider television. In television, picture and sound are transmitted by a carrier wave that propagates much as light does. This wave cannot be broadcast directly around the curvature of Earth; hence the signal from a television station has only a limited range. To serve a large area, a television network must set up relay stations at strategic points so that the signal from the original transmitter can be amplified and re-transmitted. Today such relay stations are on the ground; tomorrow they will be in the sky. It is not at all hard to bounce

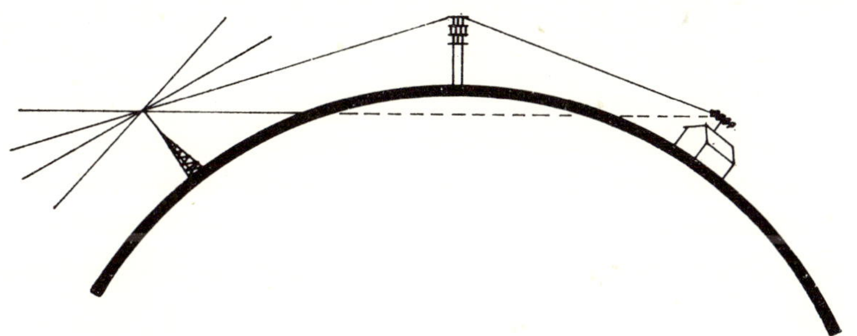

a signal off a satellite in such a way as to transmit it to an area far around the curvature of the planet.

Intercontinental television service will be arranged by means of a string of satellites placed in strategic orbits. In fact, a satellite can be made to hover continuously within sight of a given location: the trick will be to place it in an orbit at a distance of 22,300 miles from Earth; in this set-up the revolution is made in twenty-four hours. Assuming the satellite is on an easterly course above the equator, it will appear to remain motionless in the sky, because Earth's twenty-four-hour rotation will exactly keep pace with the satellite's motion in its orbit. Several such satellites strategically planted along the equator will be, in effect, several "anchored" relay stations. They will receive a signal from the ground, relay it, and broadcast it back to Earth, thousands of miles away. A world-wide television network will thus be established.

In December 1960 an aircraft company in California and a telephone company on the East Coast announced plans for the construction of television and telephone communication satellites which will even be offered to private industry for commercial use. Cake-shaped and only about 30 inches in diameter, such a proposed relay satellite will weigh only about 32 pounds, yet it will afford direct intercontinental television service and will be able to handle the business of hundreds of transoceanic telephone lines. In future long-distance calls we shall talk from one continent to the other by first sending our voice into space and then back to the other partner halfway around the world.

Large satellites of the *Tiros* type carry television cameras, a miniature transmitter, and solar power cells to run the equipment. Signals from such an electronic eye in space can be fed into our television channels, and thus vistas of the heavens as seen from far out in space will be viewed in our living rooms. We can then watch our own planet floating serenely in space!

The television camera can be equipped to act as a telescopic scanner to spot ships and airplanes. This will bring about a new era in transoceanic traffic control. Forest fires can be spotted, and icebergs can be tracked on their trips to warmer waters. There are many things that need watching on the face of the planet, and modern civilization will derive much benefit from a satellite with keen eyes.

What about the benefits to be expected from "manned" space flight? Today we can say that man is probably space-fit. It may be difficult for him actually to live in space for days, weeks, or months, running a large, manned space station; but this might not even be necessary. Tremendous advances have been made in the development of automatic equipment which does just about everything—talented robots of science that outperform man in sensitivity, reliability, and speed of performance. A satellite with a net payload of, say, 1,000 pounds could accommodate a big load of instrumentation—probably enough to do a better job than a man, who would need at least this much weight in equipment just to stay alive.

"Putting a man in space is a stunt: the man can do no more than an instrument, in fact can do less . . ." These are the words of Dr. Vannevar Bush, war-time director of the office of scientific research and presently chairman of the Board of Governors of the Massachusetts Institute of Technology, as he testified to the House Committee on Science and Astronautics in 1960. "There are far more serious things to do than to indulge in stunts. As yet the American people do not understand the distinctions, and we in this country are prone to rush, for a time, at any new thing. I do not discard completely the value of demonstrating to the world our skills. Nor do I undervalue the effect on morale of the spectacular. But the present hullaballoo on the propaganda aspects of the program leaves me entirely cool."

Of course, it is man who builds machines in the first place and makes them perform. The instrumentation of a 1,000-pound satellite will have to be serviced and readjusted for different jobs from time to time, and this is where man comes in. An orbital space ship can carry him into space now and then to look after his equipment, which, itself, doesn't mind staying out there for years.

A trip into space will be at least as tiring as a long-range mission in a modern military aircraft. After thirty hours in the air, a man is more than ready for a good night's sleep. In space flight, things will at best be no easier. For a long time, manned space flight will probably be practical only in the form of missions lasting a day at a time. But a day is ample for a round trip to a satellite in need of a human hand and a mind that can make the right decision.

When the tools of space flight have become well developed, we can tackle the most important and most promising project of the space frontier: the control of weather and climate. Control means far more than the observation of weather by satellites. Taming the unruly atmosphere will be one of the most significant tasks in human history.

Consider what weather has meant to man. In 1955 the unholy trio of hurricanes, "Connie," "Diane," and "Ione," killed more than 250 people and left a damage of almost two billion dollars in their wake. A few years ago merciless storms hit the dikes of Holland and broke through; cities, villages, and thousands of acres of fertile land were ruined by the salty floods. Such disasters could be listed by the hundreds; though men consider themselves masters of Earth, we all know the atmosphere is still untamed. Every single year tornadoes, hurricanes, floods, droughts, and sudden freezes demand their toll in human lives, misery, and material damage. And each day, in the course of our affairs — scientific, industrial, artistic, commercial, or just personal — each of us is at the mercy of fickle weather. There is little we can do with our meager knowledge; at best we make an effort at foretelling what the atmosphere has in store. And, be the forecasts right or wrong, the elements blow freely over civilization and wastes alike.

Winning control over the turbulent atmosphere sounds like a task for miracle men, but it could be done. Only our lack of adequate scientific knowledge prevents us from making weather to order. One of the basic rules in applied science is that there can be no control without intimate knowledge. To make further progress in weather control we must have a better understanding of how our atmosphere behaves.

The atmosphere is a vastly extended shell of gases that envelop Earth entirely. It is constantly on the move, and its actions form a world-wide, inseparable whole. Storm centers and quiet areas, jet streams and air masses—these do not recognize national boundaries; they move freely across the artificial frontiers erected by man. And despite advances in meteorology, these weather phenomena are still only spottily observed. There are not enough weather stations to keep track of the world-wide, interconnected activity of the atmosphere as a whole. A map of all weather stations of the world shows that they are heavily concentrated in civilized nations in the northern hemisphere; they thin out rapidly toward the poles, the equatorial zone, and the southern hemisphere. Regular observations are almost totally lacking over wide stretches of the oceans outside the regular lanes of sea and air traffic. Hardly anything is known about the atmosphere above the south polar cap and the ring of oceans to the north of it. We are today getting data that show only pieces, here and there, of the whole picture.

A meteorologist can collect a lot of valuable data at a particular weather station: temperature, humidity, speed and direction of wind, cloud cover, and rainfall. But his station is a mere pinpoint on the globe, giving him only a worm's-eye view of the grand scene above. If data from many stations are collected and coordinated by a central office, a somewhat blurry picture can be figured out, but a superior view is still lacking. What weather science needs is a 360-degree view of the global panorama of the atmosphere: an over-all, simultaneous view of our planet from the outside — a view from space.

The meteorologist need not be fired out into space to look for himself: satellites can do the job. The fabulous *Tiros* satellite series has already made a most promising start in this area of research. These satellites are perhaps the most useful and the most practical ones produced and operated so far. The name *Tiros* stands for "*T*elevision and *I*nfra-*R*ed *O*bservation *S*atellites." These are special designs for global weather observations. These drum-shaped instrument carriers weigh 280 pounds and are powered with 9,260 solar cells which entirely cover the round surface of the flat cylindrical shape. Each of the satellites is equipped with two television cameras, one with a wide-angle lens and the other with a telescopic lens. With a straight-down view of Earth, each picture from the wide-angle lens covers an area approximately 750 miles square, while the telescopic lens delivers enlarged pictures pin-pointing an area a hundred times smaller. The pictures are stored on video-tape and played back to the ground upon radio command from stations in California and New Jersey. In addition the *Tiros* satellites are equipped with five detectors of infra-red radiation to measure the amount of radiation given off by Earth from different parts of the ground and the atmosphere.

Between April 1, 1960, when it was put in orbit, and June 17, 1960, when the transmitter gave out, the first *Tiros* satellite delivered nearly 23,000 cloud panoramas from space, of which more than 60 per cent were excellent. Results obtained so far are extremely encouraging for improving forecasting techniques as well as for vital information in weather emergencies. At one time the weather eye in space showed cloud patterns in the Caribbean Sea that were inconsistent with a storm warning previously issued by the Weather Bureau. After a recheck of this evidence from space with conditions in the area, the storm warning could be cancelled. At another time the satellite spotted a very unusual cloud formation; two hours later a tornado appeared in this area. There can be no doubt that these miraculous helpers of meteorology will completely revolutionize weather forecasting within the next decade.

The *Tiros* satellites mark only the beginning. The big weather satellites of the future will collect vast amounts of data from all over the world – data from which the behavior of the atmosphere, in particular regions, can be predicted accurately. This will be the first phase of Operation Weather Control: making meteorology and climatology reliable sciences. Giant electronic computers, fed with the satellite-collected data, will produce long-range weather calendars, foretelling rain or shine, and warmer or cooler temperatures, for every important spot on Earth long in advance. Many important activities of our civilization are still insecure because we are still only half-guessing coming weather. The big satellites and the computers will take much of the guesswork out of agriculture and water-resources planning, travel, business, and sports and vacations. The savings in agriculture alone will pay for the big satellites many times over within a few years – and satellites are by no means cheap.

Yet this is only the beginning. As soon as man has a thorough understanding of the forces that make weather and climate, he can do something about them. Already rainmaking through the "seeding" of cloud formations has been done successfully. In time, man will find other, more efficient means to trigger the huge weather machine and make it produce what he wants. In the coming age of "planetary engineering" we shall make our weather and climate to order.

The control of weather and climate can succeed only on a global basis. The peoples of all nations will have to work together toward the common goal. And this is perhaps the most valuable result we can expect from our efforts to conquer space: the creation of powerful forces for peace and understanding between nations.

To mankind, Earth is by far the most important planet in the solar system. The glittering satellites are best used as tools of Earth research, not as weapons; they should eventually be planned, built, and run by a common effort of all the peoples of the world. As man's first triumphs in the art of space flight, they can lead to the most important goal of all: a better, safer planet for all.

9 • *Life and Planets*

The age of space travel has given us an entirely new feeling and consciousness about the true nature of our home in the universe. We raise our eyes to the sister worlds: the other planets. And the most natural, the most obvious, the most fascinating question comes to mind: Is there life on the other planets – perhaps kindred souls who, with intelligence at least as keen as ours, are conscious of the wonders of creation?

Ever since it became established that Earth is but one planet among others, people have mused about the possible existence of life "out there." In the past this question tantalized a few scientists, philosophers, and story-tellers; today it fascinates the entire world. It is this fascination that accounts for the popular excitement about "flying saucers."

Some years ago, newspapers printed reports of "a flying saucer" that made a "landing" in the desert of the Southwest. This was the story that gave birth to a character which has meanwhile become a stand-by for cartoonists all over the world: the little green man from outer space. The flying saucer, said the report, was found partly wrecked. Its form was strange, the source of its power unknown, but stranger still was the material of which it was built. This material was extremely light in weight, but very strong; it had a brilliant color and, at places where the pieces were welded together, looked like glass joined to copper. But what caught public imagination was "the crew" of this outlandish craft. They were small "humanoid" figures, only two or three feet tall, and as green as grass. There were four of them and they were all dead, presumably smothered by the gases of our atmosphere – noxious to their alien systems. These strange little people presumably hailed from the planet Venus, our nearest neighbor in space.

The story, of course, was a hoax, less credible even than accounts of the Loch Ness monster, which some scientists still consider worth investigating. But this flying-saucer story contains some significant points of psychology. People like to believe in life on other worlds and to picture the planets as peopled by intelligent beings like us. The reason for this belief lies much deeper than the current interest in space satellites and interplanetary flight. Never before has there been such a strong emotional undertow in the public's view about science. The flash of the first atomic bomb profoundly shook our hope and belief that the path of science leads straight to a better, safer life for all. Many today gaze upon science and scientists with mistrust and even fear. Where is science taking us? Never has science been mixed up so strongly with international relations. The anxieties of the cold war rise and fall with the changing feelings of national superiority or inferiority in science.

The flying-saucer myth springs from these emotions; it is an echo of the atomic bomb. Shortly after the first bomb exploded, the first flying saucer landed. And the saucers have been with us ever since. Millions of people believe that flying saucers exist and are, indeed, space ships manned by an intelligent race from a distant planet.

The saucer people are not generally looked upon as a menace. Reportedly they are peaceful; seldom have they committed a hostile act. Having traversed interplanetary space, this race must have long since created a civilization, technology, and science vastly superior to our own. Yet we can take heart despite their superiority in scientific accomplishments: the little green men are of small stature – dwarfs compared to us. Superior body size is a great compensation to anyone – to any group – with an inferiority complex.

We do, however, like the little strangers to be superior to us in civilization. This means that they have solved the problems that face Earth's inhabitants today, particularly the threat of atomic self-destruction. So hope appears to be the strongest force that keeps the flying-saucer myth alive. Wouldn't it be wonderful if this peaceful people would come in and solve our problems?

Such are some of the reasons why ordinary people – people who only a generation ago would have shrugged indifferently at the question of life on other worlds – are deeply fascinated by it today. Yet it has long been human to wonder whether there are people on the other planets, and what they might look like. Such wondering is more than just fascination with the unknown, or idle curiosity about alien races, or scientific speculation. It goes back to the deep musings in all of us about ourselves and the human role in the universe.

The question about life on other worlds began as a matter of belief, hope, and wishful thinking, not as a matter of scientific inquiry. Yet if we looked upon this question only with the eyes of a twentieth-century scientist, we might miss an important and interesting phase of our story. The idea of an inhabited universe is a dream thousands of years old, found in the religions and myths, the fairy tales and the folklore, of many peoples. It has captured the imagination of poets and story-tellers, and has occupied the thoughts of great philosophers. Even in our time the philosophical implications of this question are as powerful as the scientific ones.

Our story begins thousands of years ago when man looked upon the forces of nature and saw in them gods, demons, and spirits. The blue heavens seemed the only home worthy of powerful gods, though a few did inhabit the "underworld." The heavenly bodies themselves became divine personalities. The mighty god of the Sun was the giver of light and warmth, but also the cruel creator of heat and drought, and the dealer of death. The silvery Moon was the deity of fertility, and Venus was the goddess of love. In the mind of ancient man there was life on the stars; the stars themselves were alive – more alive than mortals on Earth, who died while the eternal stars remained the same.

The religion of the Babylonians was a star religion. Not only were the stars gods, but after death the souls of mortals joined the gods in their heavenly quarters. From these dreams it was only a short step to the ideas of the Greeks, who looked upon the stars with the curious eyes of mathematicians and thus began a science of astronomy. They recognized that Earth must be a sphere, and they realized that the Moon was not a disk fastened to the sky, but a world in itself. The Pythagoreans, and the great Greek philosophers Plato and Aristotle, assumed the existence of plants and animals on the Moon, and believed that this sister world was populated not by spirits but by people as conscious and physically alive as we are. The Moon people were a favorite subject of ancient writers; indeed, science fiction could be called the invention of the imaginative Greeks.

The followers of Plato held that the planetary spheres were both an abode and a birthplace of the souls of men. Each soul, it was believed, dwelt on that planet whose character was most compatible with the soul's own state of mental and moral perfection. And the ideas of Greek philosophy found their way into Christianity through the writings of the old church father Origen, who lived in Alexandria during the third century A.D. To Origen, Earth was but one of the lowest of worlds among millions of others that existed in the wide spaces of the heavens.

These other worlds were homesteads of the souls of men, and there these souls were purified. They ascended from one planetary station to the next until, at the greatest height, they finally reached perfection. In Origen's writings the planets were a kind of purgatory designed by the Creator to cleanse the souls step by step toward the highest state of moral and spiritual purity. The Christian concepts of heaven and hell are closely linked to these ideas, and Dante's *Divine Comedy* also expresses the ancient dream of life on worlds other than our own.

Three hundred years later Origen was declared a heretic by the church, and his writings were banned. Through the centuries that followed, the church found it unthinkable that there should exist more than one humankind in the universe. The Savior had come to *this* world to save mankind; other such worlds were not described in Holy Writ. Thomas Aquinas, the great theologian, took the emphatic stand that there was only one mankind selected to be saved from sin, and his teachings became unquestioned dogma.

It was left to the clear thinkers of the Renaissance to sweep away the mists of the Middle Ages and again open the eyes of man to the multitude of worlds in the sky. The French philosopher and scientist Pierre Gassendi, who revived the old Greek idea of the atom, also renewed speculation about an inhabited universe. He saw no conflict between these views and the stand of the church. On the contrary, he once wrote: "Let us not become guilty of sacrilegious hypocrisy in our belief that God would be unable to create on other worlds intelligent beings equal or even superior to us — beings that are cognizant of their own worlds, admiring their richness and praising the origin of all things."

The ideas of Gassendi were shared by most of the creators of modern science: Johannes Kepler, discoverer of the laws of planetary motion; the Dutchman Christian Huygens, founder of the wave theory of light; and the philosopher Immanuel Kant, the great theoretician of the rational mind. Around 1680 Huygens wrote a book in which he approached the problem of life on other worlds almost in the fashion as we would do it today. He held it probable that the major conditions for life were in existence on all planets: day and night, air, water, and land. If these conditions are met, Huygens argued, then we must expect that life has been created on all planets, including intelligent races who reason about themselves and the universe around them. He even thought that the stars have planets like the Sun's; he envisioned a universe teeming with life.

But in the nineteenth century, science lost much of the poetic and philosophic feeling it had possessed in earlier times. Many scientists became coldly reasoning judges ruling out every idea that could not be proved with absolute, mathematical precision. This was the age of the proud Frenchman Simon Pierre Laplace, who perfected Newton's mechanics of the heavens, and who proudly proclaimed that in his mathematical image of the universe there was no room for God, not even as a scientific hypothesis. This was an extreme view that few scientists shared, yet it expressed the tendency of science to concentrate on exact knowledge of material things — the kind of knowledge whose technical applications have come to dominate our modern world. By the opening of our century, there was little room in science for romantic notions about extraterrestrial life.

Twentieth-century science still tries to keep clear of human emotions and wishful thinking. The question of life on other worlds has been looked upon by most scientists with a suspicious eye. It is not a scientific question, they feel, and they are probably right. No definite proof or disproof can be given at this time. One can only guess, and scientists must avoid the habit of guessing. Hence this noble question has been left mostly to dreamers, science-fiction writers and charlatans.

With the perfection of artificial satellites and the prospect of space travel, however, we are now seeing a growing scientific interest in the possibilities of extraterrestrial life. Scientists are human enough to enjoy speculation, and there is now sufficient knowledge in astronomy, chemistry, and biology to encourage some cautious thinking on this tantalizing subject. In specific knowledge we are in a much better position than the generations before us, whose thinking remained almost pure guesswork. Although we cannot yet expect to find a direct answer to our question, there are hints to be noticed along the way.

One of the outstanding features that set a living body apart from a lump of inanimate matter is its complicated chemical structure. This complexity of structure goes back to the amazing versatility of the carbon atom. Most common materials on Earth, such as water or salt, contain several types of atoms that are joined together to form the so-called molecules. In a water molecule, for example, two atoms of hydrogen are locked together with one atom of oxygen. Sulphuric acid molecules are composed of seven atoms: two of hydrogen, four of oxygen, and one of sulphur. But the carbon atom is unique among all others. It can combine with other elements to form extremely large molecules having the structure of rings, chains, skeins, or sheets. With hydrogen, oxygen, nitrogen, and other common atoms, a great number of carbon atoms combine in different arrangements to form truly mammoth structures in the world of molecules. Many of these contain as many as several hundred thousand single atoms, joined to what is called an "organic molecule." These molecules have been called "organic" because they were first discovered in the bodies of living organisms. Albumen, amino acids, proteins, vitamins, and hormones are common examples. For a long time chemists could not produce artificially any of these organic molecules, and even to this day the complex structures of these molecular giants remain largely unknown.

Life could not be as extremely colorful, varied, and efficient as it is without this basic variability and complexity of its building blocks. Given but a few ceramic tiles of different colors, an artist could produce only a small number of simple compositions. Life is an artist that works with a set of many thousands of pieces of all colors, sizes, and shapes, thanks mainly to carbon. Without carbon or some other equally versatile element, our planet would likely be a desolate and monotonous desert, with nature repeating the same rocks, mountains, oceans, ice wastes, and clouds again and again. Carbon, more than any other substance, makes the lively, diversified world of life that we know.

The variety of terrestrial life forms overwhelms our understanding. Life is found everywhere — in the sea, in the jungle, in the air, in the arctic, in the desert. Its forms range from whales to primitive organisms so small that they can be seen only under the most powerful microscopes. Life is extremely resourceful in its ability to adapt itself to all sorts of environments. As a result of millions of years of development, the plants and animals of Earth have been able to make homes virtually everywhere in the surface layers of this planet.

The tremendous flexibility of life is one of the chief arguments advanced in favor of the existence of life on other planets. Given enough time, could not life — whatever it may be — find forms that would flourish under just about any conditions?

Unfortunately, despite their great resourcefulness, living organisms have their limitations, the most severe of these having to do with heat. A living organism can exist only within an extremely narrow range of temperatures. A temperature approaching that of boiling water will kill any form of terrestrial life. This is why boiling is the best method of sterilizing: surgical instruments are absolutely free of germs and bacteria after being dipped in boiling water. A few highly specialized animals and plants live in the pools that form around hot geysers; they can live in water at about 160°F. – water that would scald human skin badly. But these heat-loving creatures mark the upper limit of temperatures at which life on Earth can exist. Temperatures in excess of 120°F. are fatal to most living things.

The temperature range for life is similarly restricted at the lower end. Higher forms of life are killed by low temperatures. Some microscopic forms can survive at very low temperatures, but are then reduced to a completely inactive, dormant state. During the cold winters of the arctic, all land life except man that is able to survive lies dormant. It does not propagate; it is completely inactive; it just barely exists. Even the processes that lead to decay, being themselves life processes, are effectively stopped by low temperatures. That is why we keep some foods in the deep freeze.

Active and fruitful life can exist only in an environment of moderate temperatures. The building blocks of life – the giant organic molecules – are very delicate and have little resistance to heat, and practically all the complex organic compounds upon which they depend are likewise destroyed by heat. If we boil milk to sterilize it, vitamin molecules are destroyed and the nutritional value reduced. Pasteurization of milk is done by raising its temperature to about 160°F.; this kills germs but leaves the somewhat less sensitive vitamin molecules intact.

Low temperatures are less detrimental to the complex structure of organic molecules. But the power of chilled molecules to react with each other is impaired. In a living organism, very complicated reactions must go on all the time. If they are slowed down, life itself slackens and finally comes to a standstill.

Temperature rules the world of living things. The most profound changes take place in marine life; for example, when the shift of an ocean current changes water temperatures by only a few degrees, whole populations of marine creatures perish, and others may multiply. But temperature restrictions become more and more stringent as life develops higher forms, such as the birds and mammals. These creatures have won some degree of independence of environmental temperatures by developing the feature of warmbloodedness. A temperature-control system in the body is certainly one of the most marvelous developments of nature. Without built-in thermostats, warm-blooded animals, including man, would not have such keen senses, such fast reactions, such endurance, such efficient nervous systems, and — not least — such brains. Consider that the living matter of the human race has been kept, since its beginnings, at a temperature of just about 98.6°F.! This consistency shows in itself how important is the maintenance of proper temperature in the existence of higher forms of life.

So when we set out on our search for life in the deep spaces of the universe, we must look for matter that exists at moderate temperatures. But we find that this requirement is rarely met in nature: practically all the matter of the universe exists at extremes of temperature. About half of it is closely packed into the gaseous bodies of stars, which are at temperatures of thousands and even millions of degrees. The other half of the matter of the universe is spread out thinly as the single atoms and tiny dust particles of interstellar space. The temperature of these dust motes is extremely low: about 450°F. below zero. But a tiny fraction of cosmic matter has been concentrated into those rare bodies we call planets. These bodies, drastically different from the great bulk of universal material, circle around stars which, in some instances, keep their surfaces between the temperature extremes. Here, then, the trail through space begins to get "warm"; for the creation of planets is the first great step toward the creation of life.

Assuming that the chemistry of life must be basically the same throughout the universe, when we come to sister worlds we can measure the chances of the existence of life with a thermometer.

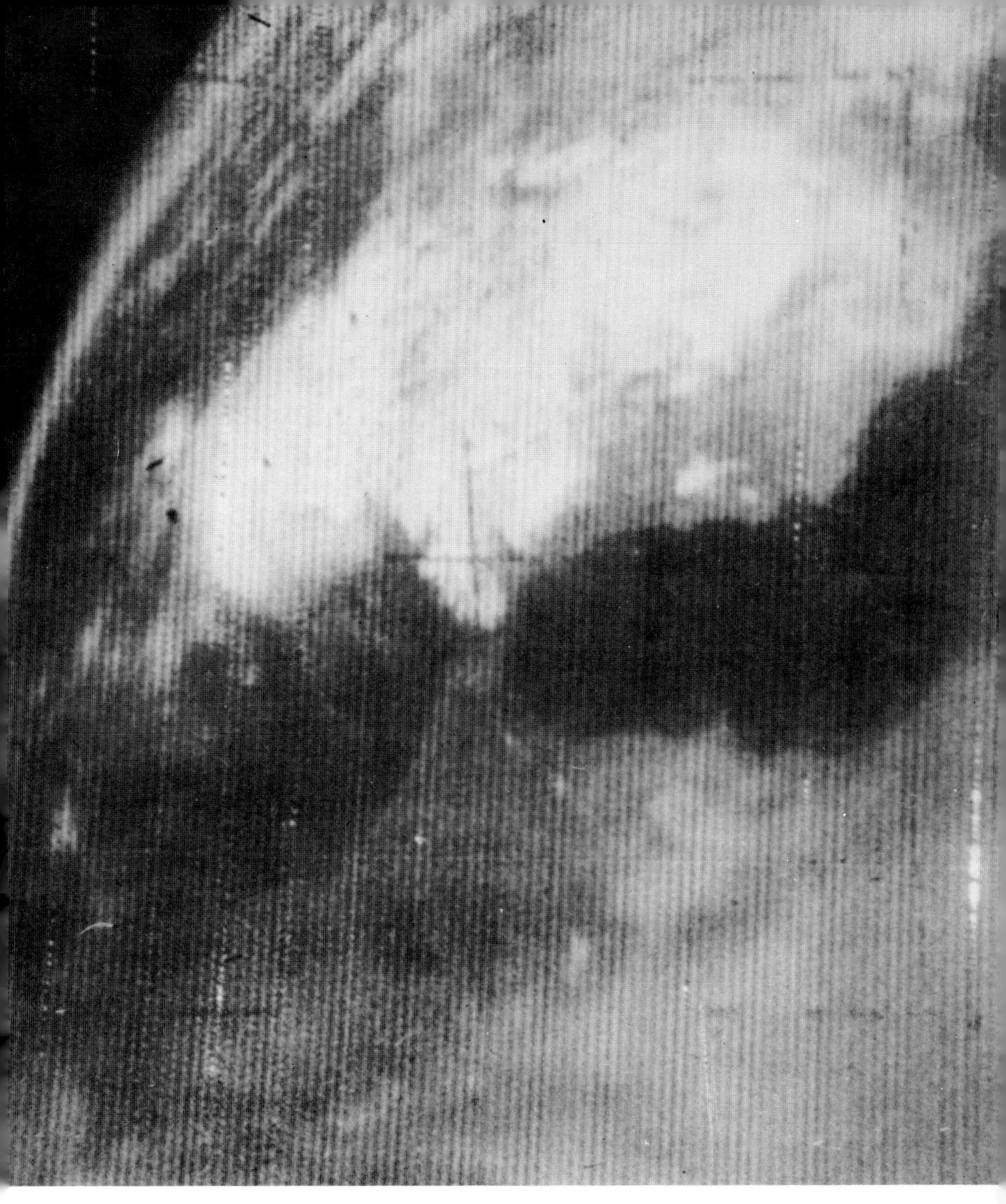

A Tiros weather satellite picture of the Straits of Gibraltar showing the Mediterranean (black area to right) and Atlantic (black area to left). The white cloud cover that shows up over Portugal and most of Spain gives observers the kind of information they need to predict coming weather conditions

Before pictures like this were available, people thought life might exist on the Moon. Now we know that these stark mountains and awesome craters do not harbor life. Soon, perhaps, a visiting astronaut from Earth may upset this age-old state of affairs

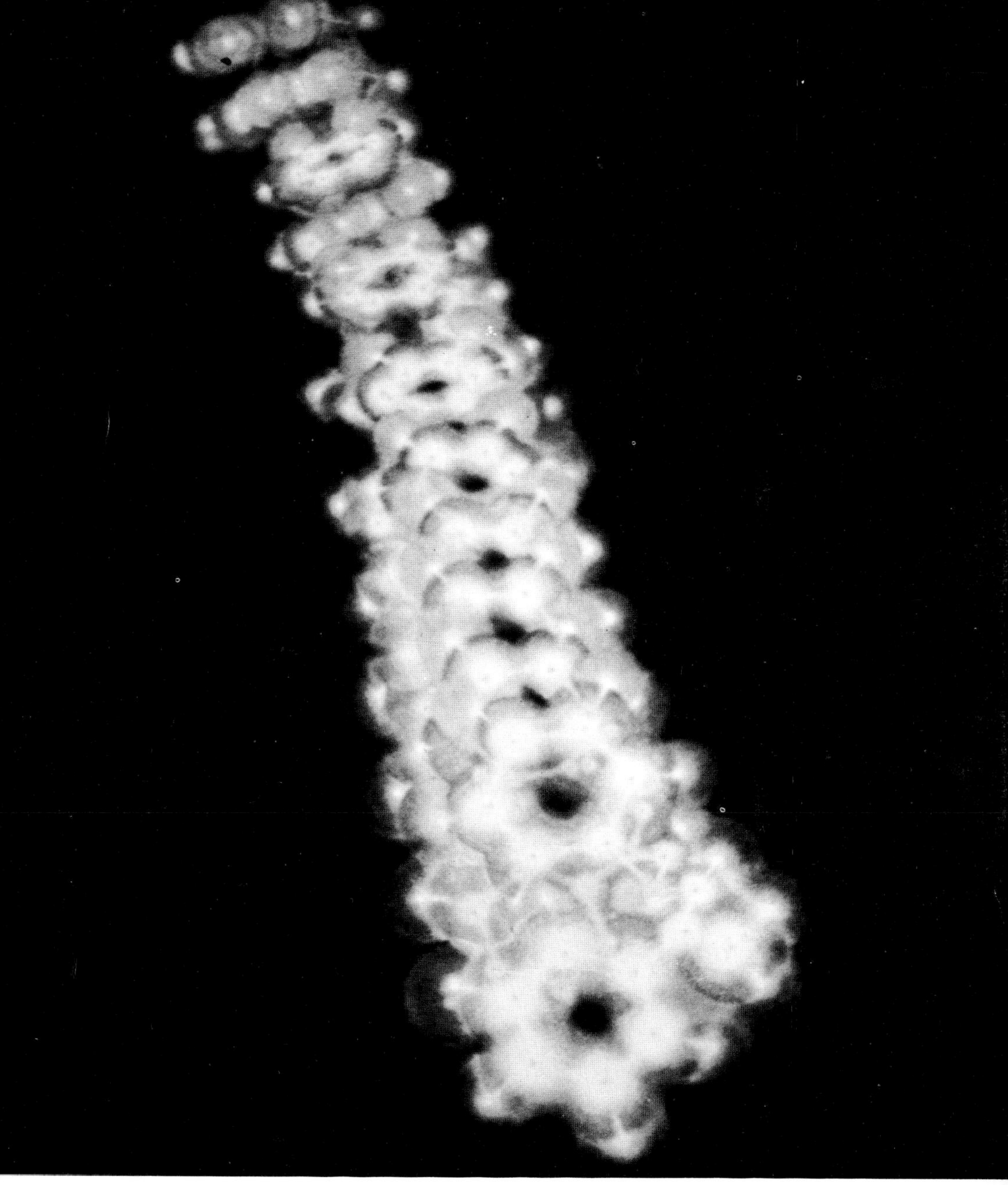

This model of a polystyrole molecule demonstrates that the world of the microcosm is no less complex than that of the macrocosm

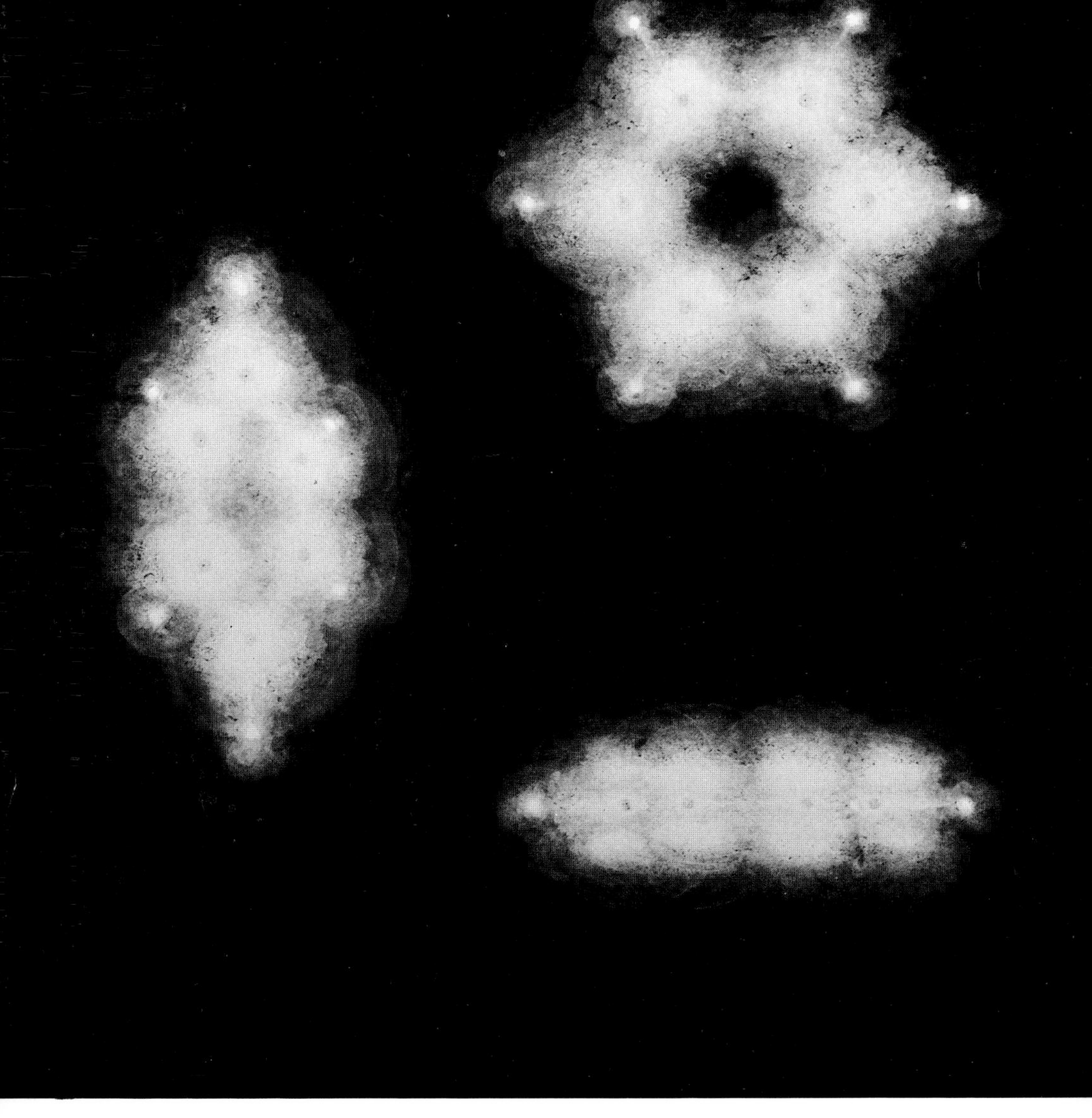

Three views of a benzene molecule. Although it looks deceptively simple, this molecule is at the base of a number of substances including drugs, plastics, dyes and explosives

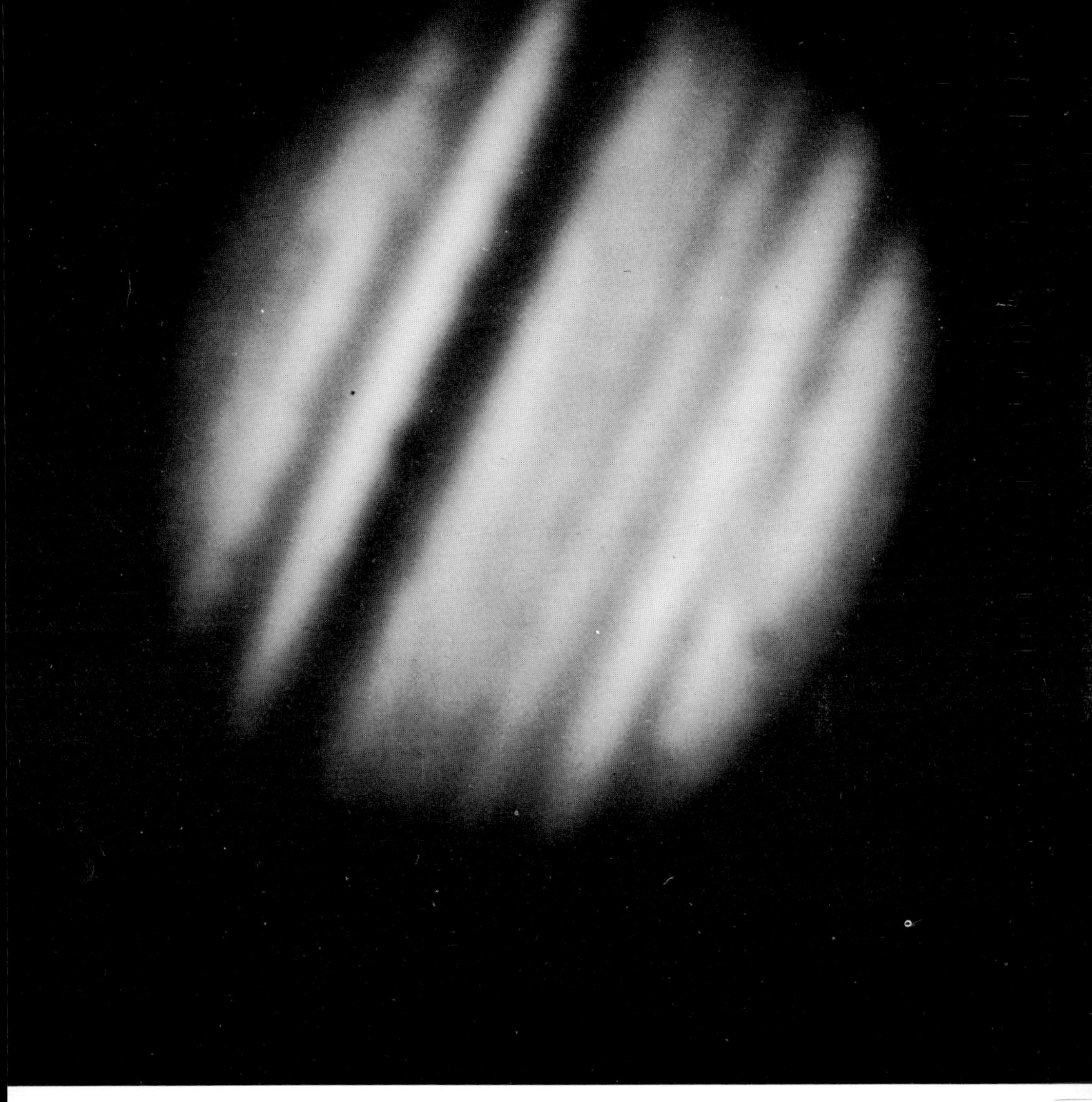

The largest of our Sun's planets, Jupiter circles far out in space, well beyond the range of life-giving warmth

Although, like Jupiter, Saturn is too far from the Sun, and thus too cold, for life, space probes might throw some light on the composition of the planet's beautiful but mysterious rings

Mars is, perhaps, the most intriguing of our sister planets. A tip of white shows seasonally at each pole, leading scientists to suspect the presence of water, forming polar ice caps

To our eyes the famous Pleiades look like a compact cluster of stars. The individual stars, however, are from two to four light years apart

Thin clouds of dust and gas extend through space, in some cases becoming dense enough to cut off light from more distant sources. A well-known example is the "Horsehead" nebula in the constellation Orion

The extent of the universe is vast and man can see only so far into it. This cluster of nebulae is in the constellation Como Berenices. At or near the limit of the 200-inch telescope, its distance was estimated at between 500 and 1,000 million light years. More recent calculations indicate it may be twice that distance

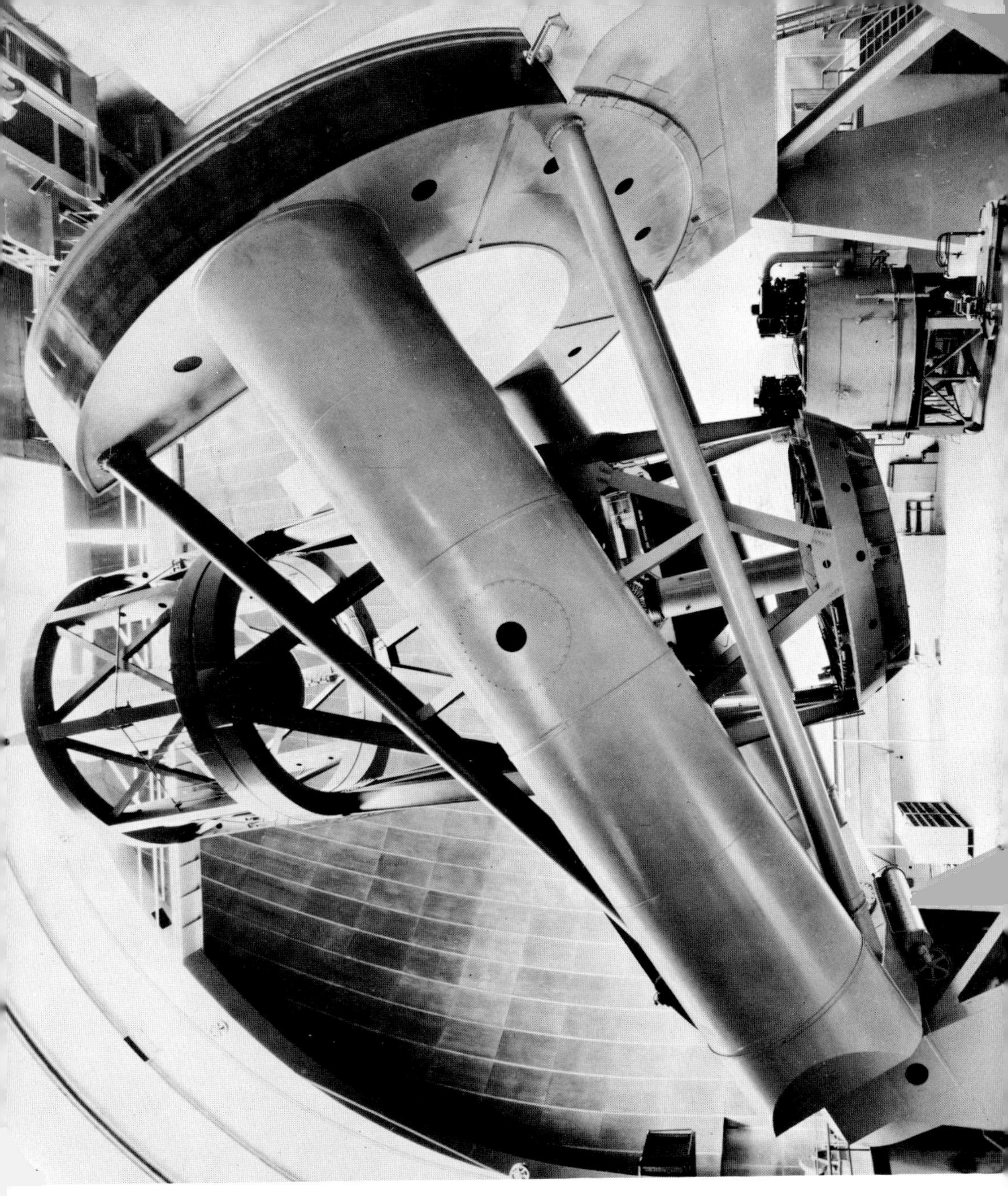

The 200-inch Hale telescope points upward from Palomar Mountain in California. Much that we didn't know before has been discovered by this powerful "spyglass"

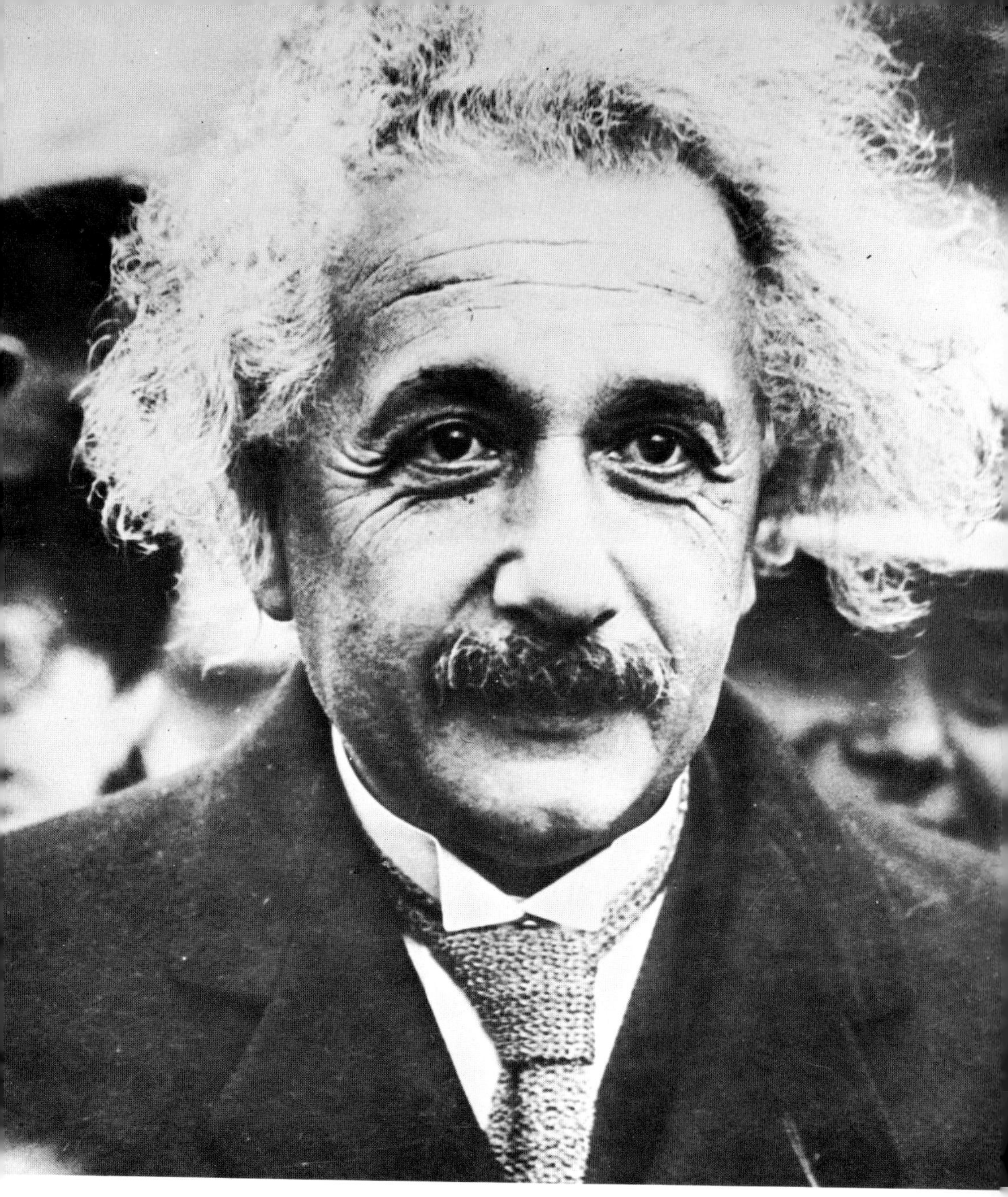

The late physicist Albert Einstein as photographed in 1942 when he was in his sixties. Einstein's theories of relativity revolutionized man's concept of the universe

Britain's Jodrell Bank radio-telescope picks up radio waves emitted by galaxies. Able to "see" farther than optical telescopes, this instrument gives data on location, age and condition of celestial phenomena. Scientists hope to gain more exact knowledge from this and from the bigger radio-telescope now being built in West Virginia

The 48-inch Schmidt telescope in California is not outmoded by its bigger companions. One of its more recent uses has been in the study of intergalactic matter

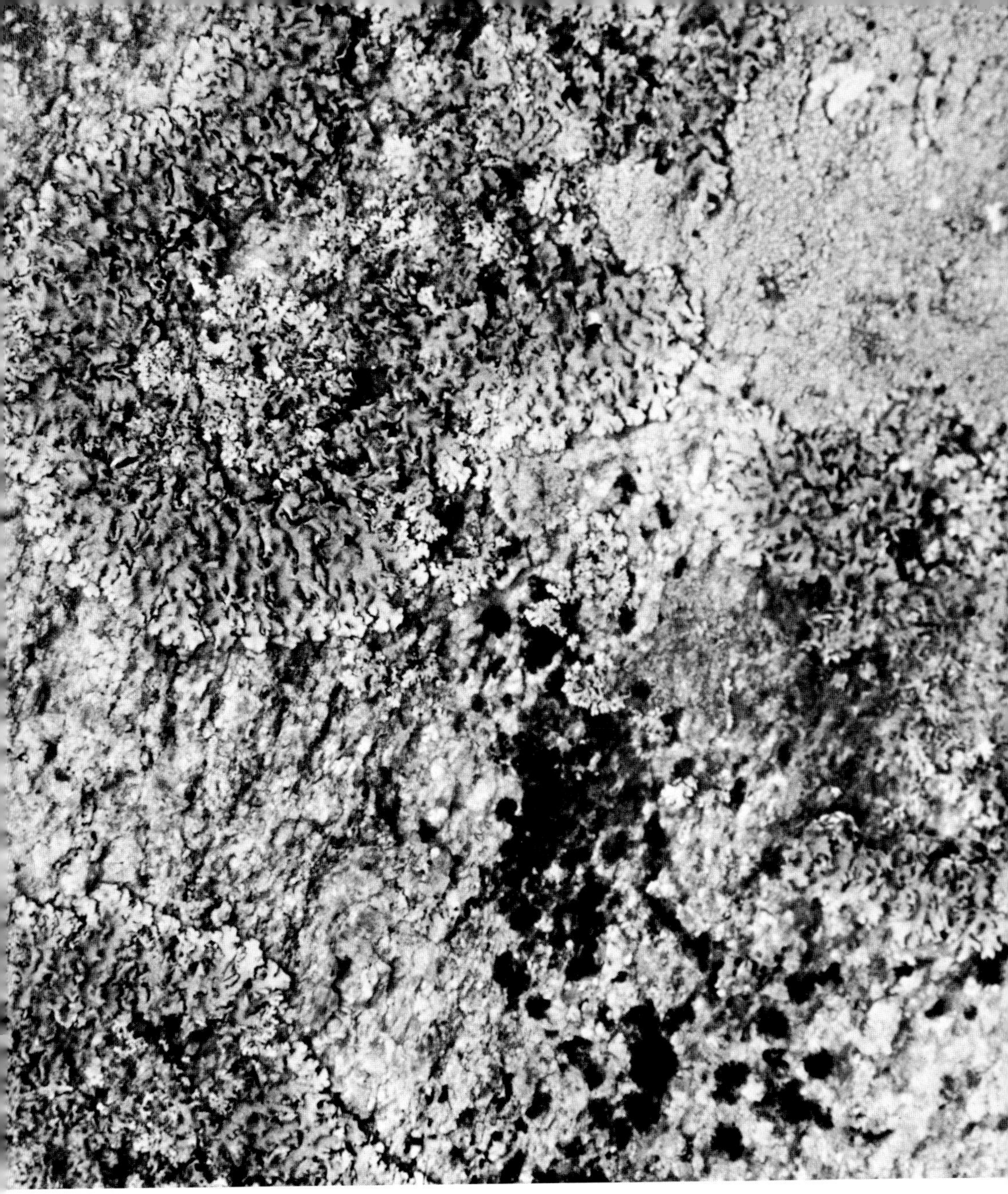

This lichen is an example of the kind of life — if any — that is likely to be found on Mars. Lichens have the ability to store oxygen and thus can survive where other life would "suffocate"

An Army Redstone missile blasts off, taking Alan B. Shepard, Jr. on his historic sub-orbital flight into space. This marked the first time an American has left his home planet and an auspicious beginning for America's Project Mercury program

Our own planet is really a standard: it is the ideal place for life to flourish. Our distance from the Sun is such that most of the globe gets enough, but not too much heat, for life processes. Most of Earth's surface, land and sea as well as the lower levels of the atmosphere, lies constantly within the narrow temperature band of life. Any other world must, so far as we know, offer similar conditions if it is to sustain plants and animals comparable to Earth's.

Conditions are quite different on the Moon. Here there is no atmosphere to speak of. In daylight the Sun's rays beat fiercely upon the bare rocks and dry dust of the rugged landscape; and at night the temperature plunges down to around 200°F. below zero. The Moon lacks the bulk necessary to hold an atmosphere and, therefore, any open water.

Even if through some miracle we could endow the Moon with an atmosphere like Earth's, and fill the hollows of its surface with water to create oceans, the Moon would quickly lose both. The movements of the atoms and molecules of the lunar air would, like those of Earth's air, become greatly speeded up by the Sun's heat. Some racing particles in the topmost layers would get enough of a boost from collisions with other racing particles to escape gravity and shoot off into interplanetary space. This process occurs in Earth's atmosphere, too, but Earth's gravity is six times as strong as the Moon's; hence the Moon would lose its atmosphere much faster. Within a few thousand years the entire lunar atmosphere would be gone, and with it the long-since-evaporated lunar oceans. The Moon, like the planet Mercury, simply does not have the necessary gravity to hold a mantle of air.

To say that the Moon has no atmosphere is not strictly correct. On the surface of the Moon there are probably a few forlorn particles of heavy, slow-moving gases that could be called an "atmosphere." But these rare wandering particles would hardly deserve the name of "air."

It is difficult for us, who spend our lives at the bottom of a huge sea of air, to imagine touring an airless world. When the first lunar explorer scoops up a handful of the fine lunar dust and tosses it upward, it will fall back to the ground like a shower of pebbles. There will be no air to keep the powder afloat. For the same reason, our explorer will find that a feather and a piece of lead, when dropped, will fall at the same speed – but not as fast as the lead would fall on Earth, with its stronger gravity. On the Moon our space man will perform still another amazing trick: he will throw a soft paper handkerchief a mile! There will be no air resistance to stop it, and, once it is on its way, little gravity to pull it down.

Lacking water and air, the Moon lacks a means of moderating the extremes of temperature between day and night. On Earth the atmosphere and the oceans store the heat of the day and dissipate it slowly during the cool night; thus we feel neither the heat of the day nor the cold of the night at its worst. But on the airless Moon the noonday temperature rises beyond that of boiling water, then drops to below 200°F. below zero after sunset. For billions of years, it seems, the Moon has undergone a temperature variation of about 400°F. in its regular monthly cycle.

No bit of life similar to any living thing known on Earth could exist under the rigorous lunar conditions. It is highly unlikely than any living organism could survive either the awful heat or the awful cold; and it is still more unlikely that any organism could survive both, by turns. It seems positively certain that our satellite is a lifeless world, every inch a desert.

The Moon is much smaller than Earth and all the other planets. What of the possibilities of life on these other, larger worlds?

Consider the innermost of the planets: Mercury. Although this has a diameter one and a half times greater than the Moon's, its mass like the Moon's is too slight to have a gravity strong enough to hold an atmosphere and surface water. Moreover, because Mercury is three times as close to the Sun as we (and the Moon) are, the solar heat on its surface is very great – enough, in fact, to melt tin. We refer, of course, to the side of Mercury that faces the Sun: the other side is correspondingly cold. For Mercury is, by any standard, a place of extremes. Telescopic observations indicate that Mercury does not rotate about its axis as Earth does. One side of Mercury always faces the Sun, just as our Moon presents to Earth one and the same side at all times. Because Mercury does not rotate, one side is unendingly roasted, and the other unendingly faces the lightless depths of space. And Mercury has no atmosphere to carry heat from the hot side to the dark side. So the dark side must have a temperature approaching absolute zero: around 450°F. below zero.

Mercury, then, is even more forbidding to life than the Moon is. Only with artificial protection could any living organism survive on this planet. And that life could ever have developed under such adverse temperature conditions is unthinkable. Mercury is a world devoid of life, another wasted sphere of glowing – and frozen – rock.

The next planet in line is Venus, the lovely morning and evening "star." Its distance from the Sun is about two thirds of Earth's. In size, Venus is a true sister of Earth: its diameter is only about 4 per cent smaller. The solid sphere of Venus is always hidden behind a dense white veil that never lifts. For a long time it was believed that this thick overcast was composed of clouds like those familiar to Earth-dwellers.

Venus has often been compared to the Earth of past geological ages. The much greater solar heat on Venus has been thought capable of producing vast jungles nourished by hot, moist air and almost everlasting rains. The entire surface of Venus could be covered by vast marshes and swamps, populated by huge saurians like those of Earth's great Age of Reptiles. But such thoughts probably are pure romancing. Spectroscopic study of the upper atmosphere of Venus has detected little water vapor and no oxygen. The air of Venus contains great amounts of carbon dioxide, which plants might use. But the presence of the gas probably causes lethal temperatures at the surface of the planet. Although the temperature at the top of the overcast has been measured at about 0°F., in the deeper layers the gases are much warmer. Carbon dioxide traps solar heat as the glass panes of a greenhouse do. The deepest layers to which our measurements can reach are at around 125°F., and this implies a temperature on the ground of at least 300°F. – hot enough to make any known form of life impossible. Venus just cannot be the home of any small green people!

Skipping Mars for the moment, let us consider the outer planets of the solar system: giant Jupiter, beautiful ringed Saturn, green Uranus and Neptune, and distant Pluto. These members of the solar family can be ruled out as potential homesteads for life with one sweep: they are much too cold. Circling the Sun at enormous distances, far removed from life-giving warmth, they remain forever encased in their heavy armor of ice and frozen poisonous gases, methane and ammonia. Life could never unfold on any of these frozen worlds. Higher forms would die instantly in the freezing cold; even the cold-resistant species of bacteria and spores would be doomed to complete inactivity before they, too, succumbed.

No wonder, then, that we pin our hopes on Mars, among all the known worlds in the sky, as the most likely abode of life. Today the red planet is the focus of the age-old dream. It has been the focus, in fact, since an astronomical event that occurred eighty-odd years ago.

The year was 1877. The art of telescope making was well advanced, though telescopes were not as big as they are today.

Before the 1870's, astronomers had concentrated mostly on the motions of the heavenly bodies; now they were turning more to the physical nature of the stars and planets. The Milanese astronomer Giovanni Schiaparelli was interested in the planets particularly, and one night he made an exciting observation. On the reddish disk of the planet Mars he saw, or believed he saw, a criss-cross pattern of extremely fine straight lines connecting the darker areas. The network was visible only at times when the turbulence of Earth's atmosphere, always an exasperating handicap to astronomers, was at the minimum. Schiaparelli called these fine lines *canali,* the Italian word for "channels," or "canals." He called them *canali* just as the dark spots on the Moon had been called *maria,* "seas."

The suggestive word *canali* electrified the world. "Canals" suggested something created not by nature but by people. The "canals of Mars" started a world-wide orgy of speculation about the possible existence of Martians that has not subsided even to this day.

Around the turn of the century an American astronomer, Percival Lowell, became the champion of this idea. He urged the building of the now well-known Lowell Observatory near Flagstaff, Arizona, for planetary study and devoted his life to the study of Mars. A long series of observations seemed to support his original, imaginative ideas. He noticed wide regions on Mars that seemed to change their color and shape in rhythm with the seasons on that planet. In winter these regions would show a yellowish tint; this turned green or bluish green in the spring, and later in the year the color would change to brown, then back to yellowish. It seemed that as a polar cap melted with the onset of spring, the canals grew more distinct. In fall and winter, when the polar cap reformed, the canals would almost vanish.

In 1907 Lowell published a book that became famous: *Mars as the Abode of Life*. Here he described his observations, offering the explanation that Mars was inhabited by intelligent people who were fighting a brave battle against the odds of dwindling water resources. The Martians, he thought, had criss-crossed their planet with a gigantic, planet-wide irrigation system. Melting snows from the polar caps were their only source of water. Each year the canals directed this meager but precious flow toward the warmer regions, where it was used to water the fields.

Lowell's ideas were enthusiastically received by the public and by a number of his colleagues, but most astronomers were highly skeptical. Of course, no proof or disproof was forthcoming; none was possible. Lowell's theories had to be considered as practically pure speculation until more hard facts were at hand. Conclusive facts probably will not be known until the day when the first space ship lands on Mars.

After Lowell's death, in 1916, climatic and atmospheric conditions on Mars became more accurately known through the use of more highly refined instruments used with the telescopes. The new data made Lowell's Martians look very imaginary indeed. The air of Mars was found to be devoid of any detectable trace of oxygen, without which it seemed impossible for higher forms of life either to survive or to originate. The Martian climate, too, is hostile to life. During the night, the temperature drops to almost 100°F. below zero, and only around noon near the equator does it rise above freezing – for an hour or two. No animal on Earth could tolerate such tremendous daily variations of temperature. The conditions on Mars are so rigorous that most speculations about Martians are viewed skeptically by geologists.

The problem of life on other worlds has generally been regarded as belonging to astronomy, and scientific books and articles about the subject have been written by astronomers, almost without exception. An exceptional book, written by the biologist Hubertus Strughold, is *The Red and Green Planet,* "A

Physiological Study of the Possibility of Life on Mars." Its scientific approach has brought careful thinkers to a new outlook about Martian life. Strughold states that even very primitive animals could not exist on Mars, because there is no oxygen to support their metabolism. Oxygen is the element of life; its absence spells death. Plants, too, need oxygen to breathe, and the Martian atmosphere does not contain nearly enough oxygen to fulfill the demands of the plant kingdom. Plants, however, can perform a very special kind of chemical survival trick beyond the power of animals; photosynthesis. Dr. Strughold has shown that this trick could get around the forbidding conditions on Mars.

Photosynthesis is the remarkable life process that takes place through the action of chlorophyll, a complex chemical substance with a vivid green color. The raw materials of this process are water and carbon dioxide, and the motive power is sunshine. Plants absorb the energy of sunlight, and under the control of chlorophyll the molecules of water and carbon dioxide are split into their component elements: hydrogen, oxygen, and carbon. These are recombined to form the so-called carbohydrates; that is, sugar, starch, and fat. These products are used by the plants to build their bodies; there is an excess of oxygen, which is released into the air. The atmosphere of Earth is actually a vast storehouse of free oxygen produced by plants all through geological history. Even though large masses of oxygen are constantly consumed, the chemical activity of the plant kingdom has kept the oxygen store of the air on an even level.

Like animals, plants need oxygen for their metabolism, and on Earth there is enough of this life-giving gas in the air to go around for all things alive. On Mars there is no free oxygen in the air, and it looks as though even plants could not exist there because they need free oxygen to breathe all the time. But at this point Strughold steps in with an interesting idea. He does not hold it probable that higher forms of plant life can exist on Mars, because it has been shown that these kinds of plants wither and die if kept in an oxygen-free atmosphere. In addition, these plants would be killed by the very low temperatures of the Martian night. Strughold looks instead for the lowest kinds of plants that are particularly well adapted to stand very chilly temperatures, and he studies how they function in an atmosphere completely devoid of oxygen.

His plants are the lichens, those gray-green patches of vegetation that cling to bare rocks on the high mountains of Earth. Lichens are a sort of combination of two types of plants: fungi and algae. The fungus provides the actual body of the lichen and lends its spongy structure to the collection and storage of moisture; the algae contain chlorophyll and carry on the process of photosynthesis, producing carbohydrates and oxygen. But what is most interesting in lichens is that the body material is equipped with innumerable cells that are filled with air. The air inside these tiny pockets contains a considerable amount of oxygen: in fact, there is more oxygen inside the plant than in a comparative sample of outside air. These lichens contain, so to speak, a private atmosphere of their own in which they store the oxygen produced by the algae. And so Strughold concludes that lichens would indeed be able to survive in the air of Mars, which apparently has no oxygen.

All the necessary conditions for lichens to flourish are found on Mars. There is sunshine for photosynthesis, and both carbon dioxide and water have been shown to exist on the red planet. The oxygen required by the lichens for their breathing is produced by the plants themselves. But the oxygen is not released into the air; the lichens could not afford to waste this precious gas. Rather, they store it inside their bodies.

During the Martian night there is no sunshine to power the process of photosynthesis, and the meager stores of oxygen produced during the day would not be sufficient to carry the plant through the long night – were it not for the extreme cold. The freezing temperatures that would kill any other plant actually help the lichens in their struggle for survival. During night the plants are chilled into complete dormancy, and all life processes come almost to a standstill. Then the plants need hardly any oxygen. A few hours later, the warmth of the morning wakes the plants into activity again and photosynthesis begins anew, providing the oxygen they need to breathe.

Dr. Strughold has supported his ideas by experimental evidence at the Aero-Space Medical Center near San Antonio, Texas. He has built what he called a "marsarium" (a term corresponding to "terrarium") consisting of an air-tight glass box which contains an oxygen-less atmosphere similar to that believed to exist on Mars. The marsarium is kept in a dark room and is periodically illuminated by lamps simulating solar radiation equivalent to that on Mars. Simultaneously with the illumination the temperature of the marsarium is cycled artificially in successive periods of heating and refrigeration simulating day and night on Mars. In this marsarium Dr. Strughold has kept a few samples of lichens which he collected himself from lava beds in New Mexico. These remarkable plants have to this date not only survived this rigorous treatment for more than two years but have even multiplied and grown considerably.

Dr. Strughold's views about life on Mars probably represent the true state of affairs. They are in keeping with the known facts of biology. Living things have a way of adjusting to the conditions of their environment and making the best of them. Here on Earth life is directed outward in free exchange with the friendly atmosphere; life on Mars is directed inward, shutting out the hostile environment. Mars was never destined to be the homestead of higher forms of life; that the red planet can accommodate any life at all is remarkable enough.

Thus, even counting Mars, only one of the nine planets in the solar system is endowed with conditions favorable for diversified, flourishing life forms. That any planet should exist with the attributes of ours is amazing indeed: a position with respect to the Sun which gives it moderate temperatures; an atmosphere rich in oxygen and carbon dioxide; and, above all, an abundant supply of water in liquid form. Judging from what we know of planets generally, the odds are very much against the combination of these favorable features in one planet.

Science is often the enemy of poetic dreaming. The known facts simply do not permit us to believe in a Mars teeming with life in all its colorful variety: flowers, trees, fish, birds, and people of one sort or another. If we ever get to Mars, we shall perhaps find at best a primitive, ugly-looking growth in patches on desolate desert soil. But that alone would mean excitement for the scientist. Even lowly lichens possess the spark of life. Even the lowliest form of life is so full of secrets that it will take science a long time yet to unravel them completely – if this is possible. There is a greater gap between a lifeless piece of rock and the most primitive living cell than there is between that cell and a human being. Science, no less than the poet and the preacher, knows how marvelous the phenomenon of life really is and the actual discovery of the marvelous spark on another world will be more soul-stirring and portentous than the wildest dreams of the generations before us.

10 • *Planets Are Like Seeds*

Among the nine planets which circle the Sun only one is the abode of an intelligent form of life. But our Sun is only one among hundreds of billions in the countless galaxies. Nature is so lavish that many suns must possess planets. Could not some of these planets be other earths inhabited by other humankinds? And is it not possible that some day there will be an encounter between the men of Earth and one of these races that come from a different sun?

Some years ago a science-fiction author wrote an interstellar adventure in which the peoples of Earth planned an expedition into space to discover another intelligent race. They were determined to make an all-out effort to find a race that shared with them the awareness of creation. The expedition was to travel far beyond the confines of our solar system, where it had been established that only Earth was the home of intelligent life.

The story took place in the distant future when nearly all the difficulties of interstellar travel had been solved. The ship was a machine representing a most ingenious solution to all the technical problems to be encountered during the voyage. But the crew still consisted of normal flesh-and-blood people. Human limitations were the same as they are today; man's life span was still essentially the same three score and ten. This limited span of life became the main theme of the story, since it takes a long time for an interstellar ship to cross the endless reaches of space from our Sun to other stars of the Milky Way.

The speed of light, 186,000 miles per second, is so great that it takes a ray of light little more than a second to bridge the distance between Earth and the Moon. This light reaches the Sun after 8½ minutes, and about 6 hours later it passes the orbit of Pluto, the outpost of our planetary system. Beyond Pluto our ray of light plunges into the immense distances of interstellar space, and even though it hurtles on with undiminished speed, it has to travel more than four years before it reaches the nearest stellar neighbor of the Sun.

The fictional expedition did not aim for the nearest star; its goal was a star many times farther away, which could be reached only after several hundred years. To solve this difficulty the planners of the expedition hit upon a fantastic scheme. They collected a hundred or so young men and women who were willing to spend the rest of their lives in the ship. When these volunteers died their children would take over; but the children, too, would die before the end of the trip, as would even their grandchildren! Generation after generation, imprisoned in a forlorn box of metal, would continue to drift through space while the goal of this monstrous trip would remain visible ahead as only a cold, unblinking point of light. Its light would not grow perceptibly stronger during the whole life of any of the prisoners.

This idea might make a good story, but such an expedition would be a superhuman — no, it would be an inhuman undertaking. How could any man pass on to his children and grandchildren such a grotesque task from which they could never escape? And what would they find upon reaching their goal — that last generation of prisoners who had never known the blue skies of Earth whence their kind came? At the end of this cruel expedition what would be their chance of finding another earth? Are there other blue and green planets in the far reaches of the universe to harbor life like that of Earth?

There are millions of suns similar to our own in the Milky Way galaxy alone — and other galaxies are spread out by the millions as far as our most powerful telescopes can penetrate. But these alien suns are so far removed from us that we cannot discover any of their planets by direct observation. All we can do is to inquire into the great forces and events that once led to the formation of our own planetary system, and attempt thereby to find out how common planets may be.

Until recently most astronomers believed that the planets originated in the wake of a cosmic catastrophe. According to this idea, the planets were formed from hot gases pulled out of the Sun by a second star that almost collided with it. The average distance between stars, however, is so great that a close encounter between any two of them would be such a rare event that we could count on the existence of only two planetary systems in the entire Milky Way — our own and the one attached to the star that once grazed the Sun.

Today we know that the chemical make-up of the Sun is so much different from that of the planets that we must abandon the theory of the mating stars. Ninety-nine per cent of the Sun's mass consists of hydrogen and helium, whereas the planets contain very little of these light gases; they are built up mostly of the heavier elements such as oxygen, silicon, and the metals. It would be impossible to create Earth simply by scooping a big chunk of material from the Sun and letting it cool.

Solar material appears to be typical of cosmic matter generally. All the stars and the thin gaseous matter which is spread throughout space in the galaxies consist mostly of hydrogen and helium, with only a sprinkling of the heavy elements which form the planets and the bodies of living creatures. How then did the planets come into being out of this mass of seemingly unsuitable cosmic material?

During the formation of the solar system some process must have acted to select the heavier elements and amass them into the cold planetary bodies. This could not have been a catastrophic event; rather there must have been a slow process of evolution that gave birth to the Sun and the planets at the same time.

When the Sun was still in the process of being formed, the gas masses of the original solar atmosphere must have formed a disk as large as the entire solar system is today. This disk contained several hundred times more material than is now found in the bodies of the planets; most of the material, however, was hydrogen and helium with as little as one per cent of heavier elements. This disk was cool, hardly warmed by the young Sun. The heavier elements in this fairly dense, cold cloud collected by chemical affinity and formed numerous compounds that condensed into little grains mixed with crystals of snow and ammonia. For millions of years this chaotic material wheeled around the Sun in countless eddies and orbits, and slowly formed concentrations of denser material in the maelstrom. Gravity drew more and more material into these denser spots which were later to become the planets. Finally the Sun contracted to a smaller sphere that grew hotter and hotter and began to radiate an intense light. The light grew so strong that it exerted a pressure upon the tenuous gases in the planetary clouds and swept them out from the solar system into the vast space beyond. The larger particles of heavy material remained in their orbits. Gravity drew these particles together and in this way the planets were born. Since the time of their creation these planets have continued circling around the Sun — massive balls composed of that rare and precious stuff, the heavy elements that are but a trace in the cosmic mixture of hydrogen and helium.

This theory seems to suggest that every *single* sun in the universe must be adorned with a family of planets, created as the suns themselves were born. But most cases of stellar evolution have probably led to the formation of double or multiple stars, which cannot be expected to have planets except in rare cases. Probably only a few stars out of each hundred have planets. Yet nature is so tremendously lavish that there must be millions of planetary systems in our galaxy alone, and our galaxy is only one among millions of equals. The number of planets in the vastness of the universe must be uncountable – in fact, the formation of millions of planetary systems is likely going on even today in the never-ending processes of creation. Many of the alien planetary systems must have collapsed because of inherent instabilities; a system in which all bodies attract each other requires a delicate balance for stability. But so many planetary systems have been formed that millions must have been able to survive against the odds of chance. It appears that the chances for life to come into being must be extremely favorable, since we must assume the existence of countless planets in the universe.

But planets are like seeds, and the inexorable laws governing life and planets are very much alike. Out of millions, nature selects only a few to reach fruition. There are thousands of ways in which a planet and a seed can come to naught. A seed might fall on dry ground, it might be eaten, it might freeze, or it might decay into nothingness long before it has had a chance to germinate. Yet each autumn so many seeds are created that each following spring the Earth becomes verdant again from the seeds that survive. In the same way a planet has only a very small chance ever to bear life; more probably it will forever remain a sterile world. Millions of the planets in the galaxies must be barren because they carry no air or are without water. Many are so close to their hot suns that they are covered with oceans of molten rock. Others are so far removed from their suns that they are eternally wrapped in ice. Many are without atmospheres because their gravitational forces are too weak to hold a mantle of gases. Other millions, no longer spinning about their axes, are divided into hemispheres of burning heat and bitter cold.

The odds for a planet being wasted are very great. But life and planets alike come into being in such vast numbers that nature can well afford to abandon almost all of them shortly after they have been created. Waste is only a human concept. In utter abandon nature scatters immense resources of matter and energy to form countless blazing suns. Millions of them are virgin and lonely, and other millions are surrounded by dead planets of all sizes and varieties. Yet, among thousands of wasted ones, a few come to fruition. There must be many planets upon which a friendly sun shines — planets that carry the elements of life, water, oxygen, and carbon dioxide within their mild shells of gas and liquid. Some of these planets may carry only a few green patches of primitive plants similar to the modest life that may be existing on Mars. But many others must carry a colorful variety of living species. We can safely assume that some of these planets are the home of intelligent races. Most certainly there are alien beings who look into the starry skies of their planets and wonder. Some may still be primitives who barely understand the universe around them. But others may have civilizations vastly superior to our own. These advanced beings would look upon *us* as primitives!

Now we can understand the yearning of the people who planned that monstrous trip in the story. Man is a social animal and he has a basic urge to communicate with his equals. Then, since we must believe that we are not the only intelligent race in the vastness of the universe, could we not some day set out to visit our alien brothers who live on their planets as we live on ours? We already know that we shall find no intelligent beings on any of the eight other planets of our own solar system; we look in vain for congenial minds in our realm of the universe. But could we not reach out into the vast space of the galaxies where we would find innumerable other planets?

We have seen what immense gaps of empty space separate the stars from each other. In the "neighborhood" of the Sun the stars are spaced at several times the distance light can travel during one year – about six trillion miles! Light is the fastest of all things and probably we can never travel at this fantastic speed. Even if we had the power to propel a ship to very great speeds, the velocity of light would be inaccessible, for the free space between the stars is not entirely empty. It is filled with an extremely thin, tenuous cloud of dust and gas. At some places within our galaxy the density of this cloud is great enough to block the passage of light given off by the stars lying behind it. We see these places as great gaps and holes in the Milky Way. Even though this interstellar matter is extremely thin, it would prevent a ship from passing through at the speed of light. The atoms striking the racing ship would create a powerful particle radiation, and if the ship should pass through a denser area of the cloud of dust and gas at almost the speed of light it would burn up like a meteor. So we can only travel at slower speeds, and the time needed to reach another star becomes hundreds of years.

Livable planets are extremely rare, and there is hardly any chance that the closest neighbors of the Sun will have even a single one. We cannot expect to find a life-bearing planet within the range of space that could be negotiated in less than several thousand years. Even under the most favorable circumstances of travel in the distant technical future, man can hardly hope to bridge the awesome gaps that separate Earth from the few other planets that bear life.

But how about the alien races? Can we not expect a visit some time from another intelligent race that may span the enormous distances? The aliens may have a much longer life span than we do, or they may have found the secret of suspended animation. If each individual could live for ten thousand or more years, the aliens might possibly visit us. The one ship of Magellan's small fleet that first circumnavigated the Earth took almost three years to complete the trip. This was a long trip, and it consumed almost one-tenth of the average life span of a man of that time. If members of a long-lived alien race had the same strong urge for exploration that we have, they might be willing to invest a similar portion of their life span to reach us.

Such visits would be one explanation for the flying saucers. These could indeed be space ships, manned by an alien intelligent race, which are spying on us for their own good — or sinister — purposes. This explanation of the flying saucers cannot be ruled out entirely, though it is extremely improbable. The dramatic event of a visit from the stars would hardly come to pass within the lifetime of the human race. For such a visit the aliens must not only overcome the formidable obstacle of space; they must also win against the equally great odds of time. The aliens would have to find us not only at the right place but also at the right time.

Our Earth is over four billion years old. Life has existed for as long as perhaps two billion years. The age of man is between 500,000 and one million years. But only in the last few thousand years has man begun to muse about the nature of the stars; only for a few centuries has he even suspected the existence of other suns besides his own. We cannot reasonably expect the alien visit to occur precisely in the short span of time during which we may be able to understand what they are and whence they come. If ever there was such a visit (or if ever there will be one), then in all probability our race did not yet exist (or will already be extinct). We cannot expect that the aliens from the stars will find us *here* and *now;* such an unlikely event is beyond any reasonable hope. There is less than one chance in a million of such a visit happening at the right time.

To demonstrate how small the chances are of an encounter between two intelligent races from different planets, imagine that during the span of one century there exist several races of ants who live in small hills in the wastes of snow and ice of Antarctica. Each of these ant races, moreover, lives for only a few hours during the century. Now, some of these races invent powerful means of transportation that allow them to roam a small distance into the area surrounding their hills. For any two of the races to meet, they would have to originate close by and during the same few hours out of the century. There is hardly a chance that two living ants of two races would ever meet.

We are one of those short-lived races. Our anthill is the Earth, and the wasteland of the Antarctic is the endless expanse of the universe. As yet we are confined to our anthill, but we are about to advance beyond its limited area. Throughout time and space other races spend their short lives on other planets. Most of them are so very far away from us that distance alone will forever keep us apart. Some of these races were already extinct and their civilizations were crumbled to dust when giant saurians roamed Earth. Other races may not yet exist as intelligent species at this time; long before they can rise to the heights of reason and intelligence, we shall be gone.

There is no reasonable chance that we will ever meet any other races. We will hardly be able to reach them, and we can hardly expect them to visit us. The far spaces of the galaxies are teeming with life; but the life-bearing planets are separated by awesome gaps of space and time. We are lost in space—marooned in the small section of the universe assigned to us.

11 • *The Universe – Finite or Infinite?*

Until the Age of Space, astronomy was always considered the most esoteric of all sciences, but now it has suddenly become not only a subject of great public interest, but also a practical, applied science. Since the first satellites appeared, many people who never previously raised curious eyes to the stars are now watching the skies. Subjects that for years had been the domain of a few specialists living in remote observatories – subjects such as lunar motion and the physics of the planets – now fill newspaper columns and television screens.

Yet the most exciting problem of astronomy is utterly impractical; whatever the final answer to this problem may be, it will make no difference whatsoever to the practical affairs of mankind. This is the problem of the nature and structure of the universe in which we live and of which we ourselves are a part. This mystery is no doubt one of the greatest and most fundamental questions of science, one that has been asked time and again ever since man has had the leisure to think beyond the grim necessities of survival. It is the utter remoteness of the problem from human affairs that makes it so challenging; the human mind has always been fascinated by ideas that border on the metaphysical.

In one way or another almost every person comes face to face with an essentially unresolvable problem — the problem of finiteness versus infinity. Of course, it is impossible for the human mind to conceive of something that is infinitely small or infinitely large; but the concept of finiteness is also inconceivable when applied to space and time. Yet there is no third alternative; the universe is either finite or infinite. This is the dilemma about which any discussion concerning the nature of our universe necessarily revolves.

Even a child is likely to come in touch with this dilemma. In some drugstore window he might see a large display bottle on the label of which is depicted a white-bearded dwarf who is holding a smaller, similar bottle. On this bottle, of course, is the same label, which in turn shows a smaller dwarf who also holds a bottle, and on this much smaller bottle is depicted a still much smaller dwarf who is again holding a bottle. Although the art of printing has not yet advanced to such a state that the label on the bottle of the third- or fourth-generation dwarf is recognizable, the implication is that there is no end to the succession. The child studying that original bottle in the window might well ask: Where does this end? Obviously it can go on forever!

Surely, there must be an end somewhere; the human mind recoils before this awesome concept of infinity. Yet to the mind there is no escape, no end to this disquieting procession of dwarfs and bottles. Behind the smallest there lurks always the possibility of the still smaller.

This is only one end of infinity; things also can become progressively larger and larger. However large an object may be — a house, a continent, the Earth, the Sun, the solar system, the Milky Way, the system of galaxies, the universe — there is always the possibility of at least *thinking* of, if not visualizing, something that is still larger. If we picture the farthest frontiers of the universe as surrounded by a picket fence bearing the sign END OF THE UNIVERSE, the mind immediately jumps the fence and asks the question: "What is beyond?" No matter how far we push the fence we can always jump it, and the concept of infinity again confronts us.

So far we have discussed only the extension of space. But the dilemma of finiteness versus infinity is also forced upon us when we think of time. Was there ever a beginning, and if so, what went on before? Will there ever be an end to all things or will time stretch into that unthinkable, unending duration called eternity?

When applied to the universe in which we live, this double dilemma concerning space and time becomes particularly difficult. If the universe is finite in both space and time, we cannot help asking, What is beyond? What was before? What will be afterward? Yet the human mind is incapable of grasping the alternative of a universe stretching out infinitely in both space and time. A third possibility might be that space and time are both infinite, but the existence of matter is confined to a limited section of space and lasts only a finite length of time. This thought is no more satisfying than the other possibilities, because it reduces the phenomenon of the material universe to an insignificant event lost in the vast oceans of space and time. It appears utterly futile even to think about such mysteries because the normal pattern of our thinking is such that we cannot conceive of any solution. We have an instinctive feeling that no proof or disproof can be found to decide the issue. But human nature has never been content to resign in the face of a dilemma. In earlier times the speculations were usually based on belief, religious dogma, and even taste. Everybody was free to choose a theory – perhaps a happy state of affairs, but hardly one to reveal the true nature of the universe. For millennia there were largely fruitless debates.

Then the striking advances of twentieth-century science suddenly changed the situation. Modern astronomy and theoretical physics have given us the means to attack the question from an entirely new angle, using concepts and ideas that were previously inconceivable. The theory of relativity and such vastly improved tools of astronomical observation as the 200-inch telescope and the large radio telescopes have given us the means to discuss the age-old problem of the structure of the universe in terms that go beyond personal belief and taste. This does not necessarily mean that we are any closer to understanding the nature of the universe than we were previously; but at least the problem and the possibility of its solution have become infinitely more exciting than ever before.

In the over-all view of the universe, as it is understood by modern science, we are no longer concerned with individual moons, planets, or stars; instead, the basic units of matter in the universe are the galaxies. These galaxies, formerly known as "spiral nebulae," are extraordinarily distant objects in the sky that baffled astronomers for many years. Today a typical galaxy is pictured as a vast conglomeration of stars associated to form a huge island universe equal in size to the Milky Way. In the words of the Dutch astronomer Jan H. Oort:

> Man in the past couple of centuries has been in a position like that of a lookout watching the approach of an armada of strange objects. The objects came into view first as dim fuzzy forms. As more powerful telescopes brought them closer and closer, they were identified as collections of stars, then distinguished into systems of varied shapes and types; today we can resolve the details of internal structure in many of them.

There may be as many as a hundred billion stars in a single galaxy, and billions of such galaxies have been recorded by the powerful telescopes used to penetrate the depths of space.

Interesting as these details are, they need not concern us too much in connection with the basic problem of the structure of our universe. What is most significant here is the statement that the galaxies *are* the basic units – the atoms, so to speak – of the universe.

Two important discoveries have been made about the distribution of galaxies in the universe. The first is that they appear to be evenly spread throughout all space as far as the largest telescopes can reach. There is a tendency among them to form groups and clusters, but this clustering is averaged out statistically when the universe is considered as a whole. The second discovery was made in the early twenties by V. M. Slipher, E. P. Hubble, and M. L. Humason. They found that the galaxies are moving away from each other in such a way that the speed of flight between any two of them is in proportion to their distance. This phenomenon led to the theory of the "expansion of the universe," and it is perhaps the most important and exciting astronomical discovery of a century rich in scientific surprises.

Obviously the recognition of the true nature of the galaxies and their strange flight must give much food for thought to all those who inquire into the nature of the universe. However, these momentous discoveries would be of little use without a powerful theory that could make them understandable within a larger system of thought. It is perhaps one of the most fortunate coincidences in science that the expansion of the universe was discovered by astronomers only a few years after theoretical physicists had derived the very same idea from their mathematical equations. These equations all go back to the famous Special Theory of Relativity developed by Albert Einstein in 1905 and generalized later by him in 1917.

The theory of relativity is supposed to be incomprehensible to the uninitiated; however, any discussion about the nature of the universe without recourse to the ideas of Einstein would be futile. The theory of relativity has shed an entirely new light on the concepts of space, time, and matter — the basic entities of the universe. It was soon recognized by Einstein himself as well as by contemporary mathematicians that the theory of relativity was a unique tool for attacking the age-old question about the sort of universe that is around us. Unfortunately, one of the concepts of the theory of relativity transcends the immediate powers of human visualization. These concepts involve an element that is not set up in the structure of our thinking: the fourth dimension.

Mathematically speaking, a dimension is a major extension of space. An infinitely thin line extends only in one direction — it is one-dimensional. A plane like a piece of flat paper extends in two directions; it has only length and width, if we disregard the thickness of the paper. So a plane is two-dimensional. A box extends in three directions — height, length and width — and so space has three dimensions. The two dimensions of a plane form a right angle, as can be immediately seen from every corner of a sheet of paper. The same is true for the three dimensions of space: the three edges that are joined at a corner of the box form three right angles.

By going from our piece of paper to the box we added one dimension, moving from two to three. But here the matter ends so far as the normal limits of our thinking is concerned. It is impossible to visualize the addition of another dimension to the box. There is no "room" to add another edge to the corner

Schematic diagrams of the second and third dimensions.

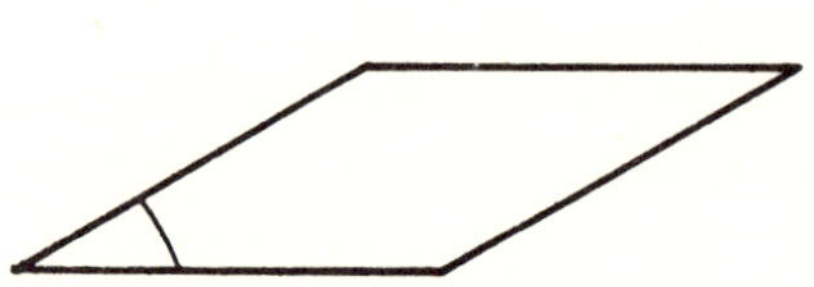

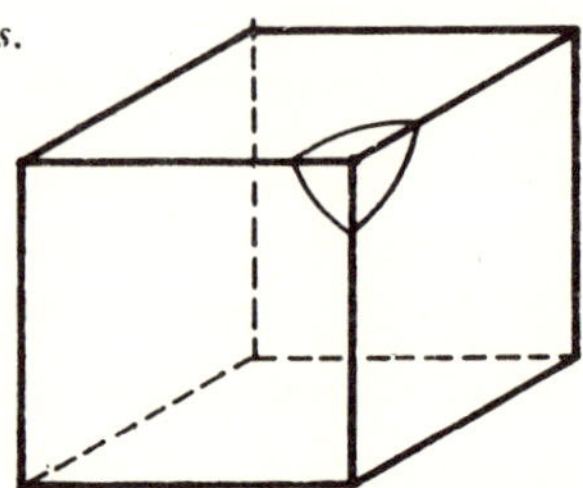

so that it also makes a right angle with the other three. In other words, the concept of a fourth dimension simply doesn't fit into the manner in which the powers of our visualization are organized.

Many believe that the fourth dimension is an extremely difficult subject, but one which only the greatest mathematicians or scientists are able to "understand" in the same way that one can "understand," or visualize, a box. This is not true. No human mind, no matter how intelligent or ingenious, can visualize a fourth dimension, because we ourselves are three-dimensional beings in our spatial configuration. Our sensory impressions and all our experiences are spread out in three dimensions only, and our thinking and the nature of our spatial conceptions must conform to our experience. There is only one way to gain knowledge about the properties of a world of four dimensions, and this is by logic and reasoning. Of course, a complete study of the strange properties of "things" of four dimensions requires the use of some fancy mathematics, but this doesn't mean that the door leading into the world of four dimensions must remain closed to all those who do not possess the key of mathematics. In fact, for the price of a little mental exercise everyone can make a journey into this strange land.

Our journey begins in a world of two dimensions from which we proceed to a world of three. When we make the transition, we shall carefully note all changes that occur. We can easily understand *and* visualize all these changes, because the world of our departure and the world of our arrival are both within the domain of our powers of visualization. There will be nothing mysterious about these changes since they are all related to our daily, three-dimensional experience. Then we shall move to the fourth dimension, and try to gain an understanding of this strange extension from a comparison with our previous observations.

We start with a world of two dimensions which may be pictured as a piece of paper that is infinitely thin. On this paper we draw a circle which, of course, is a two-dimensional configuration. Then we step into the three-dimensional world of space — the world of our experience — and we construct a sphere in this space. The sphere is a three-dimensional configuration; it relates to three dimensions exactly as the circle does to two. The two configurations are cousins, and their close kinship can be demonstrated by a simple inspection of their properties. Actually we should express this close kinship by giving both of them the same family name; we might call the circle a "subsphere." The same kinship exists between a square and a cube, and the square might be called a "subcube."

Now we are ready to make the jump into the fourth dimension, and we shall put into it a geometrical configuration that corresponds to sphere and circle. In other words, the relationship between this four-dimensional configuration and the sphere should be the same as the relationship between the sphere and the circle. There is no way to draw such a configuration, just as there is no way to represent a world of four dimensions. We can only draw a box that stands for the strange land of four dimensions and write in it the name of the four-dimensional body we put there. Since this body is also a cousin of the other two, it also gets the family name; we'll call it "supersphere." Now we must determine the properties of this supersphere.

Two dimensions can be depicted as a circle in a square, three as a sphere in a cube, but the fourth can only be symbolically depicted as space.

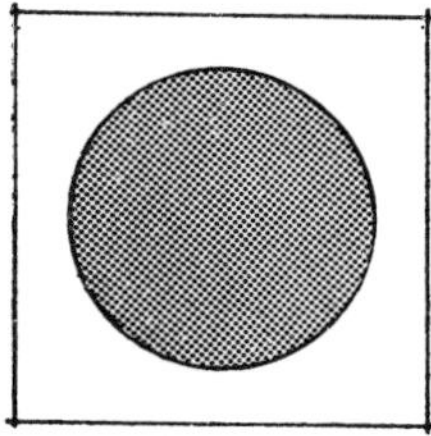

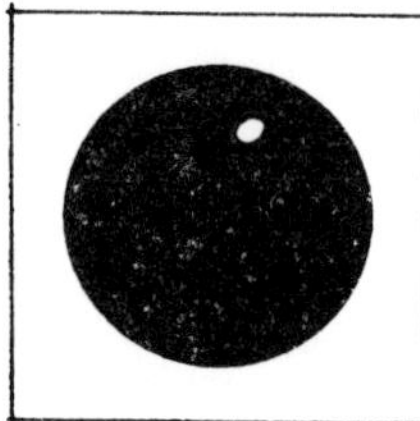

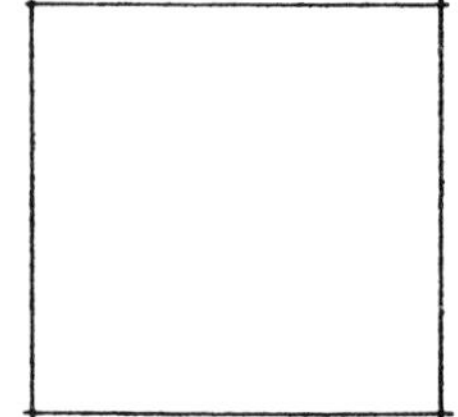

Let's go back to the circle. Its boundary is a line which itself has only one dimension. But in order to make a circle, the line requires a second dimension to curve through, otherwise it could be only a straight line and never form a circle. It is this curvature of the line that makes it possible for a circle to become a closed, two-dimensional configuration. The outer boundary of the circle, namely the line, has no beginning and no end; yet though it is unlimited, it has a certain length which can be expressed in inches. If we put a point inside the circle, it will be confined to the two-dimensional extension of the circle by the one-dimensional boundary line.

All these statements are clear and can easily be visualized. Now we shall go to the sphere and make the same statements about it. Since this transition from the circle to the sphere is accomplished by adding one dimension, we can write down all the statements about the sphere simply by adding one dimension to every part of each statement. Otherwise, there must be an exact correspondence of all expressions in the statements about the circle and the sphere:

Let's go to the sphere. Its boundary is a surface which itself has only two dimensions. But in order to make a sphere, the surface requires a third dimension to curve through, otherwise it could only be a flat plane and never form a sphere. It is this curvature of the surface that makes it possible for the sphere to become a closed, three-dimensional configuration. The outer boundary of the sphere, namely the surface, has no beginning and no end; yet though it is unlimited, it has a certain area which can be expressed in square inches. If we put a point inside the sphere, it will be confined to the three-dimensional extension of the sphere by the two-dimensional boundary surface.

These statements were written down in the manner of a translation that was made word by word without giving any thought to its meaning. They were written down without giving the sphere any thought. Simply by adding one dimension to each element where it applies, the statements about the circle have been transformed into statements about the sphere. As we now reread the statements about the sphere that we have gained in this "mechanical" fashion, we find that they are all true. They describe correctly some of the obvious properties of the sphere, and we have gained this information without so much as looking at the sphere. These statements could have been derived by a hypothetical person who lived in only two dimensions and to whom the powers to visualize a third dimension were denied. Despite this severe restriction in the structure of his thinking, he could have correctly described a sphere.

Now we are ready to describe the supersphere. Again we shall make a translation of statements. This time we shall translate the statements about the sphere, and in each case where it applies we shall add one dimension to all the expressions. The result must be a collection of statements that will describe some of the properties of the supersphere.

Let's go to the supersphere. Its boundary is a space which itself has only three dimensions. But in order to make a supersphere, the space requires a fourth dimension to curve through, otherwise it could only be a flat space and never form a supersphere. It is this curvature of the space that makes it possible for the supersphere to become a closed, four-dimensional configuration. The outer boundary of the supersphere, namely the space, has no beginning and no end; yet though it is unlimited, it has a certain volume which can be expressed in cubic inches. If we put a point inside the supersphere, it will be confined to the four-dimensional extension of the supersphere by the three-dimensional boundary space.

None of these statements appears to make any sense, and yet we must realize that they are meaningful. It is impossible to visualize a "curved space." We can picture curved lines and curved surfaces, but not curved spaces. Well, this is one of the meaningful concepts we can gain only by translation. We cannot visualize a fourth dimension, and since "the space requires a fourth dimension to curve into," a curved space is also beyond the domain of human conception. Similarly, it is a strange thing to recognize that the outer boundary of a supersphere is not something like a wall, as the name boundary suggests; rather it is a space. In our language the term "space" implies freedom of movement and is contradictory to the term "boundary." Again, we have gained a new concept by translation. A two-dimensional plane can restrict movement in the third dimension, while it affords complete freedom of movement within its two-dimensional expanse. In the same way, a space can indeed serve as a boundary restricting movement in the fourth dimension, while inside the space there is three-dimensional freedom.

When we cut a sphere in two and look at the area of the cross-section, we recognize a circle. Mathematically speaking, cutting is equivalent to reducing by one dimension. Cutting a sphere creates a subsphere, or circle. What happens when we cut a supersphere? Since cutting is equivalent to reducing by one dimension, the cross-section of a supersphere is a sphere!

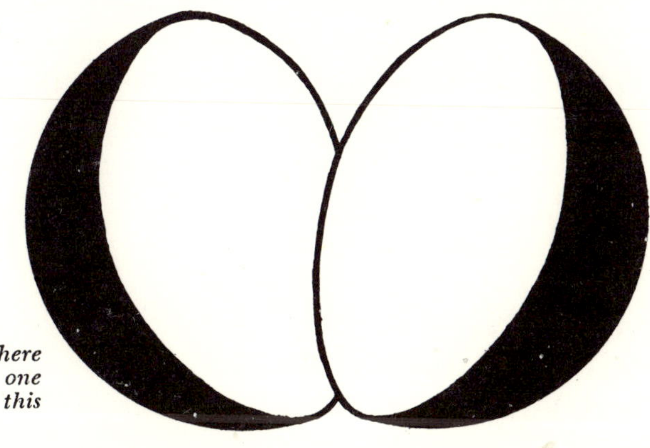

Slicing through a sphere results in the loss of one dimension giving, in this case, a circle.

Let's go back to our two-dimensional world and assume that it is populated by two-dimensional sub-people. They have no way of visualizing a third dimension, because their experiences and their way of thinking are confined to only two dimensions. You, however, are a three-dimensional being and you possess a sphere that you are going to show to these sub-people. As you hold your sphere above their world, the sub-people see absolutely nothing, since they cannot look outside their world. As your sphere touches their world, the sub-people perceive just a single point – the point of contact between the sphere and their plane world. As you now run the sphere through their world, the people see a circular disk appear. This disk grows and grows, until their world cuts across the equator of the sphere; then the disk becomes smaller again, shrinks to a point, and vanishes from their world. The sub-people would be greatly amazed by the sight, and if you should try to tell them that all they have seen was a sphere that was run through their world, they wouldn't know what you were talking about. They would admire you greatly for being able to perform such inexplicable tricks; yet to you the whole thing would be utterly simple.

Now, in comparison to the fourth dimension, we ourselves are "flat" people. If a four-dimensional superbeing possessed a supersphere and decided to pass it through our world, we would be amazed by the results. Suddenly out of nowhere, a point would appear at the place where the supersphere made

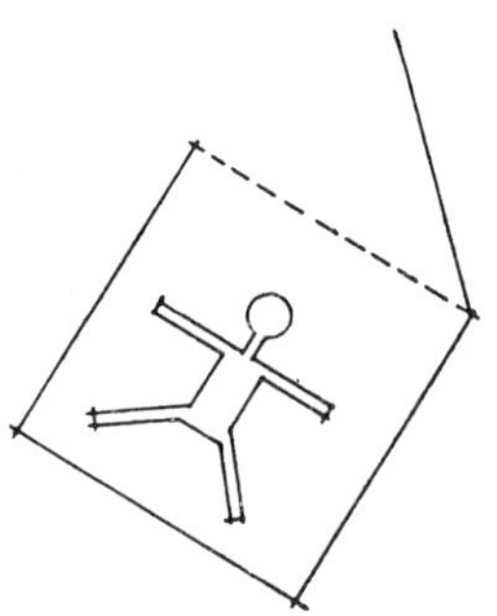

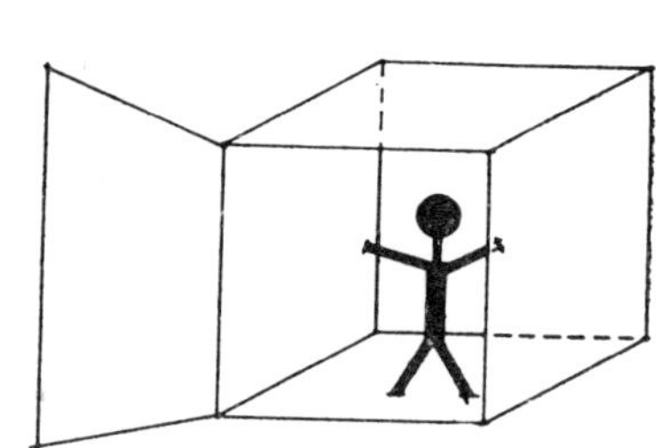

contact with our world. This point would become a sphere that would grow larger and larger; finally reaching the "equator" of the supersphere, the magic sphere would begin to shrink again until it was only a point. Then it would disappear completely from our world.

We could play another trick on our friends, the sub-people of the two-dimensional world. Suppose there is a sub-criminal who is put into jail by sub-cops. The jail, of course, is a closed square or rectangle. Once the sub-criminal is inside he is a prisoner; he cannot escape since in all directions of his two-dimensional world he finds himself hemmed in by a "wall" — the lines of the closed figure. It would be no trick at all for us to spring him without ever breaking into the jail, because to us it is wide open from above and below. We would simply grab him from above, lift him out of his prison, and whisk him away through the third dimension. To the sub-cops who would find their prisoner gone without any evidence of break-in, it would be a truly magic escape.

In the same way, to a four-dimensional super-liberator all the closed, three-dimensional boxes we build as jails are wide open from his "above" and "below." He could "lift" a prisoner out of jail without ever breaking through any of the doors or walls! A four-dimensional super-surgeon could perform truly amazing operations. The super-surgeon could simply lift a dangerous foreign object, such as a pin, from the interior of a baby's body without cutting the flesh or inflicting any pain!

All that we have found so far about superspheres and the fourth dimension seems little more than amusing mathematical theories that have nothing to do with the realities of the physical world. Indeed, the first speculations about the properties of a four-dimensional world began as a mathematical pastime. Then Einstein found that this intellectual game of the mathematicians could be most usefully applied, and he made it a part of his famous theory of relativity. In this Einstein attempted to explain the structure of the universe by means of curved spaces.

Since a curved space needs a fourth dimension to curve through, it was necessary to determine where the fourth dimension could be found in our physical universe. Einstein decided that the fourth dimension was time. Yet, space and time are two different entities. They are not immediately interchangeable in the sense that any two of the spatial dimensions, such as length and height, can be exchanged simply by turning a box on its side. Einstein applied a mathematical modification to time that made it formally equivalent to the three spatial dimensions. He called this world of four dimensions the "four-dimensional continuum" and claimed that it was indeed the universe we live in. In this concept the curvature of space was linked to the presence of matter within the universe.

There were a few inconsistencies in Einstein's first suggestion. These were cleaned up by the Russian mathematician Alexander Friedman, who made a further mathematical study of a four-dimensional continuum filled with galaxies to represent our real universe. He found by mathematical reasoning, however, that such a universe would not be stable and ought to expand like a blown-up balloon. Friedman made his discovery in 1922. It was a strange discovery indeed, because no one had ever dreamed of a universe that behaved like a balloon. But it was only two years later that Edwin Hubble of the Mount Wilson Observatory in California found evidence that the galaxies are flying apart like the fragments of an exploding shell. This discovery was immediately linked to the strange result obtained by the mathematicians who had been trying to apply Einstein's equations to the structure of the universe. The motions of the galaxies was immediately identified as observational proof of an expanding universe of the Einstein-Friedman variety.

In order to make this clear we can represent the galaxies as polka-dots on the outer surface of an expanding balloon. Since we are in one of these polka-dot galaxies, we are surrounded on all sides by others. When the balloon is blown up the other dots will all be moving away from us, and the farther away from us they are the larger their speed will be. This is the pattern of motion which was described before: since the polka-dots on the surface of the balloon are all equal in rank, they move away from each other with speeds that increase in proportion to their distances. Since the original discovery of the flight of the galaxies, measurements of the speed of recession have been obtained for the farthest frontiers of the observable universe, a distance of over six billion light years. At this distance the galaxies move away from us with a speed of 90,000 miles per second, which is almost half the speed of light. Apparently we should say that we live in an "exploding universe," rather than in one that is merely expanding.

The polka dots on the surface of the balloon served only to give us a curved, three-dimensional model of the universe. Actually, the universe is a four-dimensional continuum, and the three-dimensional space of the galaxies is only its "boundary," similar to the three-dimensional boundary of the supersphere. There is still the problem of determining the four-dimensional configuration, or structure, of the universe.

Let's reduce everything by one dimension as we did earlier, to provide a useful simile for the shape and curvature. In stepping down by one dimension, the space of the galaxies becomes a two-dimensional surface, and the galaxies themselves become flat dots on this surface. Now there are three ways to express the curvature of our "space." First, it could be flat, in which case space would have no curvature at all and would extend into infinity. Secondly, space could have a positive curvature, like the surface of a sphere; in this case it would form a closed body and would be finite. Thirdly, space could curve in the shape of a saddle as shown in the figure on page 185; this kind of curvature is negative, and the surface would again stretch into infinity since the saddle does not form a closed body. How are we to determine which case applies to our real universe?

Unfortunately, the equations derived from the theory of relativity are of little help here. These cosmological equations are not like the simple high-school equations having only one unknown, with a single solution, $x = a$. Several solutions can be derived from the cosmological equations, and painstaking studies of the behavior of the galaxies in the real universe are required to indicate what kind of universe we live in.

Each of the possible solutions of the cosmological equations is called a "model of the universe." We have already met three of them: the spherical universe which is finite, and the flat and the saddle-shaped universes which are infinite. There are other factors to consider. We must also determine the changes of the universe with time. Since the universe is unstable and apparently expanding, we must attempt to discover what happened to the universe in the past and what will happen to it in the future. Will it continue to expand or will it eventually begin to shrink?

Apparently there is a large number of possible universes from which to choose. In addition to unstable models of the universe, the Einstein equations also lead to a quite different solution. This was developed by the British scientists Fred Hoyle, Herman Bondi, and Thomas Gold. They proposed the so-called "steady-state universe," which is infinite in both space and time; more importantly, it is a universe that does not change with time but always remains essentially in the same state.

The proponents of the steady-state universe had the problem of reconciling the flight of the galaxies with their principle of a steady state. This flight tends to thin out the galactic population as the galaxies move away from each other and get lost in the depths of infinite space. According to the theory this dispersal of the galaxies is exactly counter-balanced by a spontaneous and continuous creation of hydrogen atoms in the ever-widening intergalactic spaces. This newly created matter condenses to form new galaxies to take the place of those that have vanished into infinity.

The spontaneous creation of matter in free space is perhaps not such a strange and unthinkable idea as it first sounds. It can be explained on the basis of the laws of conservation of matter and energy applied to the universe as a whole. This idea, first proposed in 1945 by the German physicist Pascual Jordan, does not contradict the known laws of nature.

There is a tremendous appeal in the theory of the steady-state universe. According to this theory, creation of matter, stars, and even galaxies has been going on since time immemorial; it still goes on today and will go on forever. The steady-state universe connects the idea of everlasting change, birth, and death with a never-changing background upon which the drama of the worlds projects itself. In this respect the steady-state universe is almost classic in its serenity, and it forms a sharp contrast to the unstable, evolving models of the universe.

Both the steady-state and the various types of the evolutionary universe are theoretically sound solutions of the cosmological equations. Theory alone cannot decide between them. In order to make a choice we must look into one special feature of the evolutionary universe that we have not yet discussed — the remote past. Since the galaxies are moving away from each other with speeds that are fairly accurately known, it is easy to compute how long ago they must all have been at the same place. If there were a motion picture of the history of the universe from its beginning and if we could run this film backward, we would see the galaxies rushing toward each other; since the farthest of them move fastest, they would all arrive at the same point at the same time. That dramatic beginning, when all the matter of the universe was condensed into one incredibly hot and compact ball, must have been about six billion years ago, according to the astronomers, physicists, and mathematicians who adhere to the evolutionary theory. They believe that the universe was created in one cataclysmic event that gave birth to all space, time, and matter.

One of the prime defenders of the evolutionary theory is the American physicist George Gamow. In the beginning, according to Gamow, the huge primordial mass of extremely hot matter became the birthplace of hydrogen and helium, which constitute 99 per cent of the mass of the universe. The disputed question of where, when, and how the heavier elements were formed has no bearing on the geometry of the universe. It took this exploding mass 250 million years to organize itself into the countless galaxies that continue to rush away from the birthplace of all things. Today the flight of these galaxies still bears witness to the cataclysmic beginning of all things six billion years ago. Will this flight motion go on forever and leave our section of space practically devoid of galaxies as they disperse themselves into the ever-increasing size of the universe? This question is still open. Since the density of matter in the observable region of the universe lies within certain limits, the force of attraction operating on this matter can be calculated. This force would act to slow the outward flight of the galaxies. The question arises as to whether this attraction could ever check the galactic motion, reverse it, and finally bring the galaxies back to the original point of creation. It now appears that the galaxies move much too fast, and their mutual attraction is much too weak to pull them together again. However, it is possible that we have made a wrong guess about the amount of matter contained in the universe. If this is much larger than our estimates indicate, the mutual attraction between the galaxies would be greater and might slow down the expansion of the universe. Observations made with the 200-inch telescope by M. Humason indicate that such a slowing down may indeed be going on.

Strange as it may seem, we can learn something about the speed with which the galaxies moved a billion years ago. The reason for this is quite simple. Many galaxies which can be reached with the largest telescopes lie about one billion light years away from us. This means that the light which reaches us from them has been on its way for one billion years; what we learn from it must apply to that ancient time. Humason found that these distant galaxies appear to move faster than they should, which means that the motion was faster one billion years ago. On the basis of this evidence we must conclude that the expansion of the universe has been slowing down noticeably during the last billion years. In fact, the rate of deceleration appears to be such that the expansion will eventually stop; then the galaxies will start on their way back. Billions of years hence they will crash together in a momentous collision that will be the death of the old and the birth of a new universe. If this is so, in the very distant future another mankind may observe another universe of galaxies and come to the conclusion that each universe is but one of the pulses of creation.

It is obvious that confirmation of this decreasing rate of expansion would tend to disprove the steady-state universe theory. The rate of expansion must always be the same in the steady-state universe.

The fact that we actually look back in time when we observe the outermost galaxies gives us another possibility for deciding between the evolutionary and the steady-state universe. The tools for this test are provided by the new science of radio astronomy. It has been shown that some galaxies are sources of extremely intense radio waves that can be recorded over distances much farther than optical telescopes can reach. Every galaxy is a radio source, but the emission of radio waves becomes extremely strong in the case of colliding galaxies. The collision between two galaxies is not at all such a dramatic event as the expression implies. The distances between the individual stars within a galaxy are so great that two galaxies can intermesh in collision without a single pair of stars coming into contact. However, each of the two galaxies contains great clouds of hydrogen gas; when such clouds meet in a head-on collision large amounts of energy are released in the form of radio waves which can be detected over tremendous distances of space and billions of years of time.

A billion years ago the galaxies were much closer to each other than they are today; consequently, collisions between two galaxies must have been more frequent than they are now. With this idea in mind the British radio astronomer Martin Ryle has been sampling the universe for sources of radio waves that could be identified as originating from colliding galaxies beyond the reach of the 200-inch telescope. He arrived at the surprising conclusion that in the far reaches of space, where what we see happened more than one billion years ago, these collisions

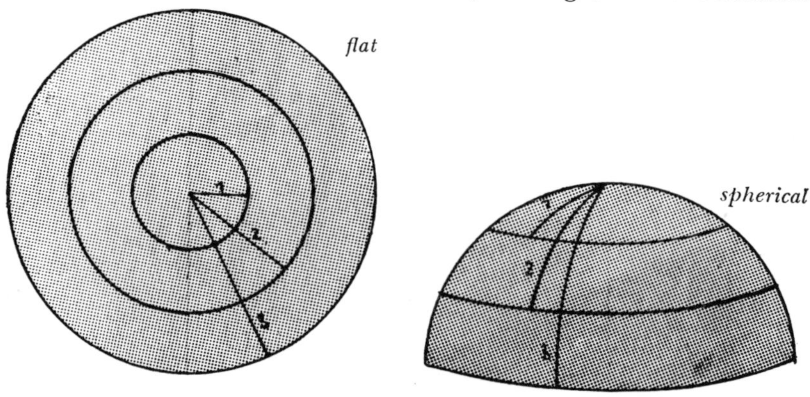

were more frequent than they are today. This conclusion also tends to disprove the theory of the steady-state universe, which presupposes that the frequency of collisions between galaxies must be the same throughout space and time.

The proponents of the steady-state universe claim that the evidence of colliding galaxies and of the slowdown of expansion is still too shaky to decide the issue. It is true that all these measurements are made at the very limit of observability, and the results must be sifted carefully from a contaminating background of possible errors; but the balance seems to swing gradually in favor of the evolutionary universe.

The steady-state universe is by definition infinite in space and time. The evolutionary universe may be either finite or infinite; if we must favor this theory, we are left with no answer to our original question about the nature of the universe. Again we have three choices: an infinite flat universe; a finite spherical universe; or an infinite saddle-shaped universe.

The previously mentioned study of Humason which indicated a slowing expansion of the universe also suggested that the universe was a kind of a closed supersphere, and that we live in a universe of limited dimension.

There is, however, another way to settle the choice between the three configurations: plane, sphere, and saddle. It can be

Three possible shapes for the universe:

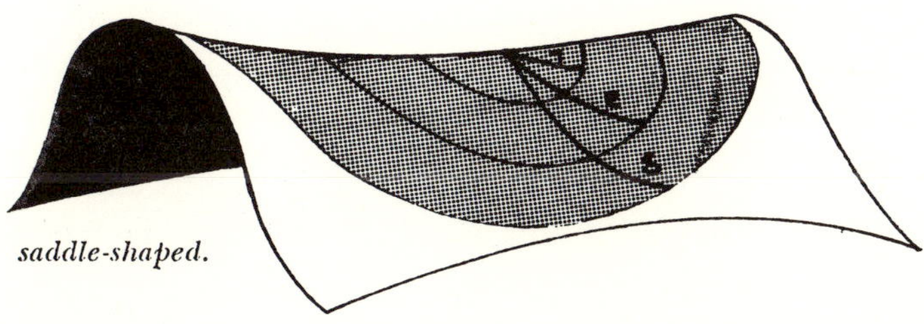

saddle-shaped.

done by simply counting galaxies. Let's consider the universe as a surface which is sprinkled with flat galaxies. Now we draw a large circle, using our galaxy as the center, and count all galaxies that fall within the circle. Then we draw a second circle with a radius twice as large as that of the first. Again we count the galaxies contained in the second circle, which, of course, includes all the galaxies already counted in the first. Next we draw a third circle, with a radius three times as great as the first, and count again. We have now three numbers, and that is all we need to find how our universe curves. We must assume that the galaxies are evenly distributed all over the universe, but this is a reasonable assumption.

Let's first take the flat universe that has no curvature. Our three circles are on a plane, and we can say that the number of galaxies contained in them must be in the proportion of one to four to nine, because the areas of the circles grow in proportion to the squares of their radii — 1^2, 2^2, and 3^2.

These simple proportions get out of kilter if the surface is curved. Positive curvature means that the circles have been drawn on the surface of a sphere, and their areas increase at a smaller rate than before. This can be shown by squashing the section of the sphere onto a flat plane. Then the surface of the spherical segment will split open at various places and leave cracks which contain no galaxies. So when the three numbers increase in a proportion less than 1:4:9, our universe must be a finite closed sphere. If the curvature is negative, our universe is shaped like a saddle, and the three numbers will increase faster than 1:4:9. This can be seen if we try to flatten the saddle to a plane. This time the surface will buckle and double up at various places so that there are more galaxies in the same area than a flat circle would contain.

Of course, the surface we have used to represent the universe lacks one dimension; but as we saw, the same conclusions would also apply to the real universe of three-dimensional space imbedded in a four-dimensional continuum. The flat world corresponds to an infinite "straight" space; the sphere corresponds to a kind of finite supersphere, and the saddle to a supersaddle of infinite extension. Our key numbers will also change, since we now have to count the galaxies that are found inside spheres of increasing size instead of inside circles. These key numbers become 1, 8, and 27, since the volumes of spheres increase as the cubes of their radii.

Such a count of the galaxies has been made, and the numbers seem to increase faster than the sequence 1, 8, and 27. This result indicates that the structure of the universe corresponds to some kind of supersaddle, which would mean that we live in a universe of infinite extension.

The fundamental questions of cosmology are far from settled, however. As a matter of fact, astronomers using the famous 200-inch Hale telescope on Palomar Mountain, California, have recently made a series of discoveries that keep the science of cosmology as fluid and uncertain – but also as fascinating – as ever. In 1960 Dr. Allan R. Sandage discovered a cluster of faint stars whose age he determined with sensitive photoelectric equipment. The stars belong to our Milky Way galaxy, and they were found to be approximately 24 billion years old. If this result stands up under repeated scrutiny, we can expect another major revision of the extension of the universe in both space and time.

Dr. Fritz Zwicky of the California Institute of Technology, using the 48-inch Schmidt telescope in addition to the 200-inch, claims to have discovered the existence of intergalactic matter. For many years astronomers have been convinced that the vast spaces between the galaxies were essentially free of matter. Now Dr. Zwicky has found masses of intergalactic gases and dust spread out within clusters of galaxies. This discovery has an important bearing upon all cosmological theories. According to the cosmological equations the average density of matter in the universe has a decisive relationship to the size, speed of expansion, and age of the universe. Dr. Zwicky believes that there may be 100 to 10,000 times more matter in the universe than was previously estimated. Should his conclusions be confirmed, a major change in cosmological thinking will be necessary.

The British mathematician and cosmologist Dr. William H. McCrea has even suggested that a decision between an evolutionary and a steady-state universe may be inherently unanswerable. He maintains that we can assert "almost nothing about what the universe is like at great distances in space and time." He believes that this view "seems more satisfactory than the recent trend toward a belief that the nature of the 'whole' universe had already been discovered." We are still uncertain about our conclusions concerning the true nature of the universe, although it appeared for a time as if the key to a solution to this fascinating problem was almost within our reach. Still, it is amazing that we are able to determine so much about it.

Should the day ever come when the facts are known and we can be certain of our conclusions, this will have absolutely no practical effect on our daily lives. Yet that final answer will probably be the most exciting discovery that science can ever make. The curiosity of man must forever find its greatest challenge in the magnificent riddle of the universe.

PICTURE CREDITS

BIS 141
Fort Worth Children's Museum 48
Professor H. Haffenrichter 131, 132
William Hubbell 143
Lick Observatory Photograph 130
Peter M. Millman 44
Mt. Wilson and Palomar Observatories
35, 36, 37, 39, 43, 133, 134, 135,
136, 137, 138, 139, 142
NASA 46, 129
Official U.S. Coast Guard Photo 40
Roy E. Swan — Minneapolis Star 33
John Stofan 34
Wide World Photo 41, 42, 45, 47, 140, 144
Yerkes Observatory Photograph 38

The publishers express their thanks to Blüchert Verlag of Hamburg, Germany, for permission to reproduce diagrams from Dr. Haber's *Lebendiges Weltall* © 1959 by Blüchert Verlag.